COGNITIVE SCIENCE COMPENDIUM
VOLUME 2

COGNITIVE SCIENCE COMPENDIUM
VOLUME 2

MIAO-KUN SUN
EDITOR

Nova Science Publishers, Inc.
New York

For permission to use material from this book please contact us:
Telephone 631-231-7269; Fax 631-231-8175
Web Site: http://www.novapublishers.com

LIBRARY OF CONGRESS CATALOGING-IN-PUBLICATION DATA

Available upon request

ISBN: 978-1-60456-775-5

Published by Nova Science Publishers, Inc. ✝ New York

CONTENTS

PREFACE

This new book brings together interdisciplinary research in memory, cognitive psychology, artificial intelligence, linguistics, philosophy, computer science, neurophysiology, neuroscience, and neuropharmacology. It covers the lastest developments and pinpoint directions for future research on cognitive sciences and theory.

In:Cognitive Science Compendium Volume 2
Editors: Miao-Kun Sun

ISBN: 978-1-60456-775-5

Chapter 1

SUBCELLULAR TRAFFICKING OF BACE1: MOLECULAR MECHANISMS IN THE CONTROL OF β-AMYLOID GENERATION

Tina Wahle and Jochen Walter *

Department of Neurology, University of Bonn, Sigmund-Freud-Str. 25, 53127 Bonn, Germany

ABSTRACT

Alzheimer's disease (AD) is the most common neurodegenerative disorder, characterized by synaptic loss, and the combined occurrence of neurofibrillary tangles and amyloid plaques in the brain. While neurofibrillary tangles are intraneuronal aggregates of the microtubuli-associated protein tau, amyloid plaques are extracellular deposits containing the amyloid peptide (A). A derives from the amyloid precursor protein (APP) by sequential proteolytic cleavage by and γ-secretase. Secretase has been identified as the aspartic protease BACE1 that cleaves APP at the N-terminus of the A domain generating a membrane-bound C-terminal fragment (CTF). CTF serves as a substrate for γ-secretase that mediates cleavage within the transmembrane domain resulting in the release of A. Because BACE1 initiates A generation, it represents a potential target molecule to decrease A production in therapeutic strategies for AD.

BACE1 is a type I membrane protein that is transported in vesicular compartments of the secretory and endocytic pathway. Consistent with the acidic pH optimum of BACE1, the secretory processing of APP occurs predominantly in compartments of the endocytic pathway. The small cytoplasmic domain of BACE1 is critical in the regulation of its subcellular transport and undergoes post-translational modifications by palmitoylation and phosphorylation. Moreover, the cytoplasmic domain interacts with proteins of the GGA family and the copper chaperone for superoxide dismutase-1. Here, we review the molecular mechanisms involved in the regulation of the subcellular trafficking and activity of BACE1, and discuss the potential involvement in the pathogenesis of AD.

* Correspondence and requests for reprints: Jochen Walter, Ph.D., Department of Neurology, University of Bonn , Sigmund-Freud-Str. 25, 53127 Bonn . Tel: + 49 228 19782; Fax: +49 228 14387; Email: Jochen.Walter@ukb.uni-bonn.de

Introduction

Alzheimer's disease (AD) is neuropathologically characterized by the occurrence of neurofibrillary tangles and β-amyloid plaques in the brain [Selkoe, 2001]. Major components of plaques are amyloid β-peptides (Aβ) that derive from the β-amyloid precursor protein (APP) by endoproteolytic processing involving sequential cleavages by β- and γ-secretase [Selkoe, 2001; Annaert et al., 2002; Walter et al., 2001]. The initial cleavage of APP by β-secretase generates a membrane-bound C-terminal fragment that contains the Aβ domain (CTFβ). This fragment represents a substrate for γ-secretase, which mediates the apparently intramembraneous cleavage of APP resulting in the liberation of Aβ. In an alternative pathway, APP can be cleaved by α-secretase within the Aβ domain thereby precluding the generation of Aβ [Selkoe, 2001; Annaert et al., 2002; Walter et al., 2001].

The proteolytic processing of APP by α-, β- and γ-secretases occurs predominantly in post-Golgi secretory and endocytic compartments, and at the cell surface [Annaert et al., 2002; Walter et al., 2001; Steiner et al., 2000]. The generation of Aβ is not only determined by the relative expression levels of APP and secretases, but also by their subcellular localization. APP as well as the secretases are integral membrane proteins and are transported in the secretory pathway from the endoplasmic reticulum (ER) to the cell surface, from where they can be internalized into endosomal/lysosomal compartments [Koo et al., 1996; Yamazaki et al., 1996, Walter et al., 1996; Walter et al., 2001; Annaert et al., 1999]. Thus, enzyme–substrate interactions could potentially occur in distinct subcellular compartments. By cell biological and biochemical experiments it was demonstrated that β-secretase cleaves wild-type APP predominantly in endosomal/lysosomal compartments after its re-internalization from the cell surface [Perez et al., 1996; Lo et al., 1994; Haass et al., 1995]. Notably, the Swedish double mutation in APP that is located directly at the cleavage site for β-secretase and causes early onset AD, significantly increases the production of Aβ by enhancing the cleavage efficiency by BACE1 in the secretory pathway [Citron et al., 1992; Haass et al., 1995; Thinakaran et al., 1996].

Physiological and Pathophysiological Functions of the β-Secretase BACE1

β-secretase was identified as the aspartic protease BACE1 (beta site APP cleaving enzyme) also called Asp-2 (aspartic protease-2) or Memapsin-2 (membrane-associated aspartic protease-2) [Vassar et al., 1999; Sinha et al., 1999; Hussain et al., 1999; Yan et al., 1999]. BACE1 together with its close homolog BACE2 form a subfamily of transmembrane aspartic proteases related to the pepsin family [Bennet et al., 2000]. While the BACE1 gene is localized on chromosome 11, that of BACE2 is localized on chromosome 21 within the critical Down's syndrome region [Saunders et al., 1999]. However, whether BACE2 contributes to the pathogenesis of Down's syndrome or AD remains unclear.

Both enzymes contain the two characteristic D(T/S)G(T/S) motifs of aspartyl proteases, which form their catalytic site [Vassar et al., 2001; Walter et al., 2001]. In contrast to all other aspartyl proteases of the pepsin family, BACE1 and BACE2 are type I transmembrane proteins with a large lumenal domain, a single transmembrane domain and a small cytoplasmic tail (Fig. 1). Although, both proteases share homology of ~ 75%, strong evidence supports a predominant role of BACE1 in the generation of Aβ.

(1) The expression of BACE1 mRNA in neurons which predominantly produce Aβ in the brain, is much higher than that of BACE2 [Sinha et al., 1999; Vassar et al., 1999; Bennett et al.; 2000].
(2) Aβ production is almost completely inhibited in mice lacking BACE1 [Luo et al., 2001; Cai et al., 2001].
(3) Antisense inhibition of BACE1, but not of BACE2, results in decreased β-secretase cleavage of APP [Vassar et al., 1999; Yan et al.; 1999].
(4) BACE1 cleaves at the β-secretase site(s) of APP much more efficiently than BACE2 [Sinha et al., 1999; Vassar et al., 1999; Yan et al.; 1999; Farzan et al., 2000].

Moreover, elevated BACE1 protein levels and enzymatic activities have been reported in the brains of AD patients [Holsinger et al., 2002; Yang et al., 2003]. However, BACE2 could also cleave APP at the β-secretase site [Farzan et al., 2000; Fluhrer et al., 2002]. Also, glial cells of BACE1 KO mice, but not of BACE1/BACE2 double KO mice could still produce Aβ, indicating that BACE2 could contribute to Aβ generation [Dominguez et al., 2005].

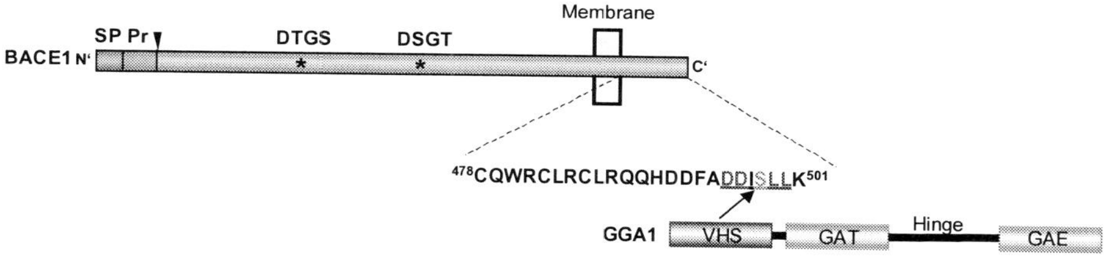

Figure 1: Schematic of BACE1. Models of BACE1 and GGA1. The aspartyl protease active site motifs (DTGS, DSGT) in the lumenal/extracellular domain of BACE1 is indicated. The amino acid sequence of the cytoplasmic domain of BACE1 is given in single letter code. The interaction motif of the BACE1 cytoplasmatic domain for GGA1 is in blue, the phosphorylation site of BACE1 (S 498) is in red. The functional domains of GGA1 are indicated. GGA1 interacts with the cytoplasmic domain of BACE1 via the VHS domain (arrow; see text for details).

Besides APP, additional candidate BACE1 substrates were identified, including the sialyltransferase ST6Gal I [Kitazume et al, 2001], the adhesion protein P-selectin glycoprotein ligand-1(PSGL-1) [Lichtenthaler et al, 2003], β-subunits of voltage gated sodium channels (Wong et al.; 2005), APP-like proteins (APLPs) [Lie et al., 2003], low density lipoprotein-receptor-related protein (LRP) [von Arnim et al, 2005] and the interleukin-1 receptor II (IL-1R2) [Kuhn et al., 2007]. Notably, BACE1 can also cleave Aβ at position 34, leading to decreased levels of Aβ40 in conditioned media of cultured cells [Fluhrer et al., 2003; Shi et al., 2003]. In line with these findings, transgenic mice with high overexpression of BACE1 showed reduced plaque load in the brain, indicating an Aβ degrading activity of BACE1 [Lee et al., 2005]. However, the biological relevance of these findings remain unclear.

Initial studies with BACE1 KO mice did not reveal an overt phenotype (Luo et al., 2001; Roberds et al, 2001). However, other BACE1 KO mouse models showed increased lethality after birth and subtle electrophysiological and behavioral alterations [Domiguez et al., 2005; Laird et al., 2005]. Recently, it has been shown that BACE1 is selectively expressed at high levels in postnatal days 0-17 correlating with the myelination period of peripheral neurons [Willem et al., 2006]. The analysis of BACE1 KO mice revealed hypomyelination and axonal bundling abnormalities, associated with accumulation of type III neuregulin-1. Alterations in myelin associated proteins and accumulation of neuregulin-1 were also observed in cortical and hippocampal neurons, indicating that BACE1 is also involved in myelination of nerve cells in the central nervous system [Hu et al., 2006]. In line with this, the activity of BACE1 has been shown to be negatively regulated by NOGO, that also associates with myelin (He et al., 2004). It has also been shown that, despite decreased Aβ levels, high overexpression of BACE1 induced overt axonal degeneration associated with thinning of the myelin sheets of CNS neurons [Rockenstein et al., 2005]. Together, these data indicate that BACE1 could play important roles in neuronal function, independent of APP processing and Aβ generation.

The Subcellular Transport of BACE1

In peripheral cells, BACE1 is mainly localized in endosomal, lysosomal and TGN compartements [Vassar et al., 1999; Sinha et al., 1999; Hussain et al., 1999; Yan et al., 1999]. The characterization of the subcellular transport of BACE1 revealed that the secretase is transported in the secretory pathway from the ER to the Golgi compartment. After translocation into the ER, the lumenal domain of BACE1 is N-glycosylated at four asparagine residues and three intramolecular disulfide bonds are formed [Haniu et al., 2000]. Mutagenesis experiments revealed a critical role of disulfide bond formation for the catalytic activity of BACE1 [Haniu et al., 2000]. Shortly after the release of BACE1 from the ER and during further transport through the Golgi, its N-terminal prodomain is endoproteolytically removed by furin and/or a furin-like protease to generate the mature enzyme [Capell et al., 2000; Bennet et al., 2001; Benjannett et al., 2001; Creemers et al., 2000]. The prodomain does not strongly inhibit enzyme activity, as one might expect for a typical zymogen, but rather plays a role in folding and forward transport of BACE1 through the secretory pathway [Benjannett et al., 2001; Shi et al., 2001]. In the Golgi compartment, BACE1 also undergoes complex modification of the N-linked glycostructures [Capell et al., 2000]. After complete maturation, the enzyme is further transported from the TGN to the cell surface from where it can be re-internalized into early endosomal compartments [Walter et al., 2001; Huse et al., 2000]. From early endosomes, BACE1 can either recycle to the cell surface, be transported to later endosomal compartments or retrieved to the TGN [Walter et al., 2001; Huse et al., 2000]. In addition, BACE1 could be directed into the lysosomal pathway as well as in the ubiquitin-proteasom system for degradation [Quing et al., 2004; Koh et al., 2005].

In polarized Madin-Darby canine kidney (MDCK) cells, the majority of BACE1 is sorted to the apical surface while APP undergoes basolateral sorting. However, a significant fraction of BACE1 is sorted basolaterally where it competes with basolaterally sorted α-secretase for cleavage of APP. Thus, Aβ production in polarized cells is determined by the differential targeting of BACE1 and its substrate APP [Capell et al., 2002].

Studies with rat hippocampal neurons revealed that BACE1 is targeted to axons and somato-dendritic compartments as well as presynaptic terminals [Capell et al., 2000]. It was proposed that BACE1 is axonally transported together with its substrate APP and γ-secretase in the same vesicles which would allow Aβ generation in these transport compartments [Kamal et al., 2000]. However, life cell imaging of primary neurons showed that APP and BACE1 carrier vesicles have distinct characteristics in the axonal transport, indicating that the two proteins could be transported in different types of vesicles [Goldsbury et al., 2006]. These data are corroborated by studies with sciatic nerves of mice showing that PS1 and BACE1 are transported separately from APP [Lazarov et al., 2005]. It will therefore be interesting to further characterize the subcellular sites of BACE1 dependent processing of APP and Aβ generation in neuronal cells in more detail.

The Role of the Cytoplasmatic Domain of BACE1 in Its Subcellular Trafficking

The cytoplasmic domain of BACE1 contains several amino acid based trafficking signals (Fig. 1). A dileucine motif has been shown to regulate endocytosis and recycling of BACE1 to the plasma membrane, and deletion of this motif increased the levels of BACE1 at the cell surface [Huse et al., 2000]. In addition, a phosphorylation site for casein kinase 1 at Ser498 has been identified that regulates the retrograde trafficking of BACE1 between endosomal compartments and the TGN (Fig. 1; [Walter et al., 2001; Pastorino et al., 2002]). While phosphorylated BACE1 is efficiently retrieved from endosomal compartments to the TGN, a mutant that can not be phosphorylated accumulated in endosomal compartments.

BACE1 can also undergo palmitoylation at three cystein residues within the cytosolic tail. Although the exact role of BACE1 palmitoylation remains to be elucidated, this modification could affect its transport and/or distribution in detergent resistant membrane microdomains (DRMs) [Benjannett et al., 2001]. DRMs might function in the trafficking of proteins in the secretory and endocytic pathways in epithelial cells and neurons, and participate in a number of important biologic functions [Simons et al., 2000]. Interestingly, BACE1 was shown to associate with DRMs where it could preferentially process APP [Ehehalt et al, 2003]. In addition, artificial GPI-anchoring of BACE1 increased the DRM association and increased β-secretory processing of APP [Cordy et al., 2003].

The cytoplasmic domain of BACE1 can also interact with the copper chaperone for copper–zinc superoxide dismutase 1 (CCS) which is involved in post-transcriptional SOD1 activation by delivering Cu^{+} to the enzyme [Angeletti et al. 2005]. The CCS and SOD1 heterodimer becomes linked via a disulfide bond, which is thought to be a prerequisite for SOD1 activation by Cu^{+} and Zn^{2+} [Wong et al. 2000]. SOD1 acts as an anti-oxidant enzyme by lowering the steady-state concentration of superoxide, and mutations are associated with familial amyotrophic lateral sclerosis [Valentine et al., 2003; Angeletti et al., 2005]. It has been suggested that BACE1 competes with SOD-1 for binding to CCS, but a physiological relevance of this interactions remains to be determined. The recently identified RNA aptamer that selectively blocks the binding of CCS by competition with the CCS binding site of the cytoplasmic domain of BACE1 might help to clarify the role of this interaction in cellular models [Rentmeister et al., 2006].

In addition to the interaction with other proteins, BACE1 can also undergo homodimerization in cultured cells and human brain tissue [Schmechel et al., 2004; Westmeyer et al., 2004]. The dimerization is mediated by the transmembrane and the cytoplasmic domains and increases BACE1 activity in cleavage of APP [Schmechel et al., 2004; Westmeyer et al., 2004].

The cytoplasmic domain of BACE1 also contains a characteristic DXXLL motif for the interaction with Golgi-localized gamma ear containing (ADP) ribosylation factor binding (GGA) proteins (Fig. 1), that mediate sorting of specific cargo proteins, like the mannose-6-phosphate receptors, from the TGN to endcsomal/lysosomal compartments [Bonifacino et al., 2004; Robinson et al, 2004]. Several groups demonstrated that GGAs also directly interact with BACE1 [He et al., 2002; von Arnim et al., 2004; He et al., 2005; Wahle et al., 2005]. Consistent with the lack of a DXXLL motif, BACE2 did not bind to GGA proteins [He et al., 2002; Wahle et al., 2005]. Notably, the interaction of GGA proteins with BACE1 is strongly increased by phosphorylation of serine residue 498 within the DISLL motif of the BACE1 cytoplasmic domain. Thus, GGAs could regulate the phosphorylation-state dependent subcellular transport of BACE1. By using dominant-negative variants and RNAi mediated downregulation of GGAs, an involvement of these adaptor proteins in the endodcytic trafficking of BACE1 has been established [He et al., 2005; Wahle et al., 2005]. Specifically, GAA1 appears to mediate the phosphorylation dependent retrograde transport of BACE1 from endosomal compartments to the TGN, while non-phosphorylated BACE1 could enter a direct recycling route to the cell surface [Wahle et al., 2005].

How could GGA 1 Dependent Trafficking of BACE1 Influence Aβ Generation ?

Several studies indicated that the secretion of APPs and Aβ was not affected by GGA proteins in cells overexpressing BACE1 [He et al., 2005; von Arnim et al., 2004, Wahle et al., 2005]. This might be partly attributable to aberrant cleavage of APP by overexpressed BACE1. Indeed, it has been shown that overexpression of BACE1 strongly increases β-secretory cleavage of APP already in early secretory compartments [Capell et al., 2000; Liu et al., 2002; Lee et al., 2005], whereas under physiological expression levels, β-secretase cleavage of wild-type APP has been localized predominantly to endosomal/lysosomal compartments [Koo et al., 1994; Haass et al., 1995; Thinakaran et al., 1996]. High overexpression of BACE1 also decreases Aβ levels in cell culture and transgenic mice by additional cleavages within Aβ [Fluhrer et al., 2002, 2003; Liu et al., 2002; Lee et al., 2005]. Thus, overexpression of BACE1 and aberrant cleavage of APP might have masked GGA-dependent effects on Aβ generation in post-Golgi secretory and endocytic compartments.

Indeed, functional analyses of cultured cells with endogenous BACE1 expression demonstrated that GGA1 is implicated in the proteolytic processing of APP. Over-expression of GGA1 reduced cleavage of APP by BACE1 as indicated by a decrease in CTFβ generation. Notably, GGA1 expression also reduced the secretion of Aβ while depletion of GGA1 by siRNA increased the generation of Aβ (Fig. 2; [Wahle et al. 2006; von Arnim et al., 2006]). Since GGA1 did not directly bind to APP, these data indicate that changes in the subcellular trafficking of BACE1 or other GGA1-dependent proteins contribute to changes in

APP processing and Aβ generation [Wahle et al., 2006]. Interestingly, expression of GGA1 also decreased the γ-secretase cleavage of APP, suggesting that GGAs may affect proteolytic processing of APP at distinct steps [von Arnim et al., 2006].

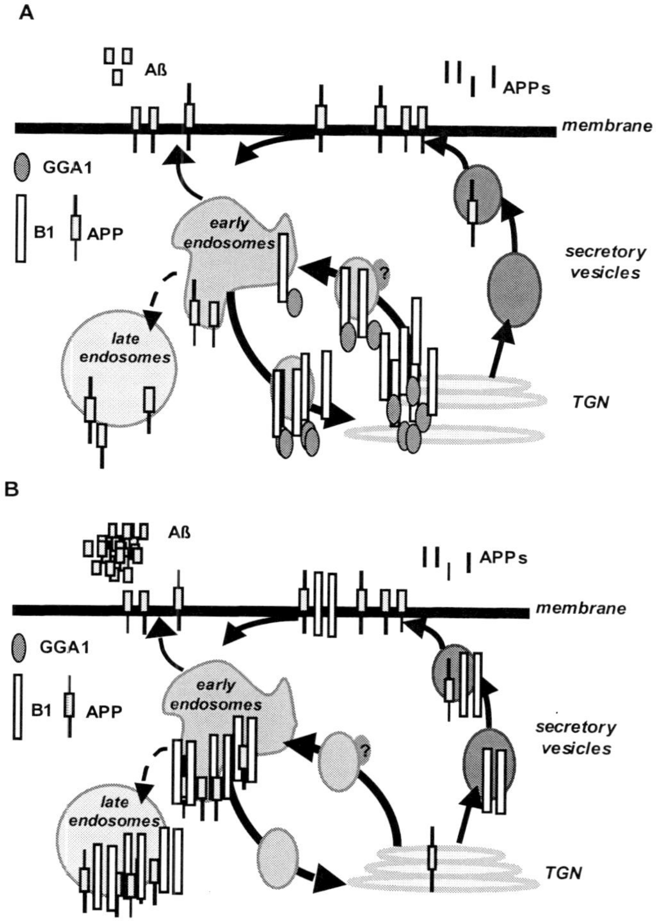

Figure 2: Model for the role of GGA1 in APP processing. Hypothetic model for the role of GGA1 in the subcellular trafficking of BACE1 and processing of APP in cells with higher (A) or lower (B) expression of GGA1. BACE1 and APP share similar transport routes within the secretory pathway to the cell surface (1), and both proteins could be endocytosed from the cell surface into endosomal compartments (2). GGA1 is involved in the retrograde transport from endosomal compartments to the TGN (3), and could thereby reduce the amount of BACE1 in these compartments. Consistent with this model, overexpression of GGA1 in cultured cells reduced processing of APP to Aβ. In addition, GGA1 might also facilitate the direct transport of BACE1 from the TGN to endosomal compartments. Decreased expression of GGA1 by RNAi led to elevated secretion of Aβ probably by increasing the amount of BACE1 in endosomal compartments. Additional GGA-dependent mechanisms, e.g., interaction with LR11/SorLA or M6P receptor (?), could also affect trafficking the proteolytic processing of APP (see text for details).

GGAs were initially identified to mediate forward transport of cargo proteins from the TGN to endosomal/lysosomal compartments [Bonifacino et al., 2004]. Whether BACE1 is also transported on this route remains to be determined. However, GGA proteins are also involved in the retrograde transport of BACE1 from endosomes to the TGN [Wahle et al., 2005]. Since BACE1 can cleave APP in both compartments to initiate Aβ generation, the GGA dependent transport between the TGN and endosomes could likely regulate the amyloidogenic processing of APP by affecting the relative contribution of α- and β-secretory processing of APP (Fig. 2).

GGA proteins could also affect the proteolytic processing of APP independent of a direct interaction with BACE1. It was demonstrated that LR11/SorLA interacts with APP and affects its subcellular localization and proteolytic processing [Andersen et al., 2006; Offe et al., 2006; Spoelgen et al., 2006]. Because LR11/SorLA also represents cargo for GGA proteins [Jacobsen et al., 2002], it is plausible that trafficking and metabolism of APP could be affected when LR11/SorLA is complexed with GGA proteins. In addition, GGA proteins are adaptor proteins that, beside cargo protein binding, also interact with the small GTPase ARF, clathrin, and additional proteins involved in vesicle assembly and targeting [Bonifacino, 2004]. Therefore, expression of GGA1 variants might also impair vesicular transport by interfering with ARF- or clathrin-dependent mechanisms that could also alter APP processing. Moreover, GGAs could affect APP metabolism indirectly by regulating the transport of other proteins. Because GGAs mediate forward trafficking of the M6PR from the TGN to endosomal compartments [Puertollano et al., 2001], and expression of this receptor has also been shown to affect Aβ generation [Mathews et al., 2002], the GGA dependent changes in M6PR trafficking could also contribute to the observed changes in APP processing.

Conclusions

In the human brain, GGA1 is preferentially expressed in neurons. Notably, total expression of GGA1 is significantly decreased in AD brain [Wahle et al., 2006]. The reasons for reduced GGA1 expression in these individuals remain to be determined, but the described findings strongly suggest that low levels of the adaptor protein may contribute to increased Aβ production which could accelerate formation of senile plaques in sporadic AD. Thus, altered GGA1 expression and activity may be a risk factor for sporadic AD.

Recently, LR11/SorLA that is also involved in the subcellular transport of APP turned out to be a risk factor for sporadic AD [Andersen et al., 2005; Rogaeva et al., 2007]. Overexpression of LR11/SorLA in neurons caused decreased generation of Aβ, whereas ablation of LR11/SorLA expression in knockout mice resulted in increased levels of Aβ in the brain, similar to the situation in AD patients [Andersen et al., 2005]. These data suggest that inherited or acquired changes in LR11/SorLA expression or function are involved in the pathogenesis of AD [Rogaeva et al., 2007]. Altered trafficking of APP or its proteases has already been shown to be an important risk factor for sporadic AD and therefore may represent an effective therapeutic approach. It will therefore be interesting to further investigate the potential of GGA proteins and other factors involved in the subcellular trafficking of APP and secretases as targets for therapeutic intervention in AD.

Acknowledgements

We thank Kai Prager and Andrea Rentmeister for the helpful discussion on the manuscript. This work has been supported by grants of the Deutsche Forschungsgemeinschaft to J.W.

References

Andersen, O. M., Reiche, J., Schmidt, V., Gotthardt, M., Spoelgen, R., Behlke, J., von Arnim, C. A., Breiderhoff, T., Jansen, P., Wu, X. et al. (2005). Neuronal sorting protein-related receptor sorLA/LR11 regulates processing of the amyloid precursor protein. *Proc Natl Acad Sci U S A* 102, 13461-6.

Angeletti, B., Waldron, K. J., Freeman, K. B., Bawagan, H., Hussain, I., Miller, C. C., Lau, K. F., Tennant, M. E., Dennison, C., Robinson, N. J. et al. (2005). BACE1 cytoplasmic domain interacts with the copper chaperone for superoxide dismutase-1 and binds copper. *J Biol Chem* 280, 17930-7.

Annaert, W. and De Strooper, B. (2002). A cell biological perspective on Alzheimer's disease. *Annu Rev Cell Dev Biol* 18, 25-51.

Annaert, W. G., Levesque, L., Craessaerts, K., Dierinck, I., Snellings, G., Westaway, D., George-Hyslop, P. S., Cordell, B., Fraser, P. and De Strooper, B. (1999). Presenilin 1 controls gamma-secretase processing of amyloid precursor protein in pre-golgi compartments of hippocampal neurons. *J Cell Biol* 147, 277-94.

Benjannet, S., Elagoz, A., Wickham, L., Mamarbachi, M., Munzer, J. S., Basak, A., Lazure, C., Cromlish, J. A., Sisodia, S., Checler, F. et al. (2001). Post-translational processing of beta-secretase (beta-amyloid-converting enzyme) and its ectodomain shedding. The pro- and transmembrane/cytosolic domains affect its cellular activity and amyloid-beta production. *J Biol Chem* 276, 10879-87.

Bennett, B. D., Babu-Khan, S., Loeloff, R., Louis, J. C., Curran, E., Citron, M. and Vassar, R. (2000). Expression analysis of BACE2 in brain and peripheral tissues. *J Biol Chem* 275, 20647-51.

Bennett, B. D., Denis, P., Haniu, M., Teplow, D. B., Kahn, S., Louis, J. C., Citron, M. and Vassar, R. (2000). A furin-like convertase mediates propeptide cleavage of BACE, the Alzheimer's beta -secretase. *J Biol Chem* 275, 37712-7.

Bonifacino, J. S. (2004). The GGA proteins: adaptors on the move. *Nat Rev Mol Cell Biol* 5, 23-32.

Cai, H., Wang, Y., McCarthy, D., Wen, H., Borchelt, D. R., Price, D. L. and Wong, P. C. (2001). BACE1 is the major beta-secretase for generation of Abeta peptides by neurons. *Nat Neurosci* 4, 233-4.

Capell, A., Meyn, L., Fluhrer, R., Teplow, D. B., Walter, J. and Haass, C. (2002). Apical sorting of beta-secretase limits amyloid beta-peptide production. *J Biol Chem* 277, 5637-43.

Capell, A., Steiner, H., Willem, M., Kaiser, H., Meyer, C., Walter, J., Lammich, S., Multhaup, G. and Haass, C. (2000). Maturation and pro-peptide cleavage of beta-secretase. *J Biol Chem* 275, 30849-54.

Citron, M., Oltersdorf, T., Haass, C., McConlogue, L., Hung, A. Y., Seubert, P., Vigo-Pelfrey, C., Lieberburg, I. and Selkoe, D. J. (1992). Mutation of the beta-amyloid precursor protein in familial Alzheimer's disease increases beta-protein production. *Nature* 360, 672-4.

Cordy, J. M., Hussain, I., Dingwall, C., Hooper, N. M. and Turner, A. J. (2003). Exclusively targeting beta-secretase to lipid rafts by GPI-anchor addition up-regulates beta-site processing of the amyloid precursor protein. *Proc Natl Acad Sci U S A* 100, 11735-40.

Creemers, J. W., Ines Dominguez, D., Plets, E., Serneels, L., Taylor, N. A., Multhaup, G., Craessaerts, K., Annaert, W. and De Strooper, B. (2001). Processing of beta-secretase by furin and other members of the proprotein convertase family. *J Biol Chem* 276, 4211-7.

Dominguez, D., Tournoy, J., Hartmann, D., Huth, T., Cryns, K., Deforce, S., Serneels, L., Camacho, I. E., Marjaux, E., Craessaerts, K. et al. (2005). Phenotypic and biochemical analyses of BACE1- and BACE2-deficient mice. *J Biol Chem* 280, 30797-806.

Ehehalt, R., Keller, P., Haass, C., Thiele, C. and Simons, K. (2003). Amyloidogenic processing of the Alzheimer beta-amyloid precursor protein depends on lipid rafts. *J Cell Biol* 160, 113-23.

Farzan, M., Schnitzler, C. E., Vasilieva, N., Leung, D. and Choe, H. (2000). BACE2, a beta - secretase homolog, cleaves at the beta site and within the amyloid-beta region of the amyloid-beta precursor protein. *Proc Natl Acad Sci U S A* 97, 9712-7.

Fluhrer, R., Capell, A., Westmeyer, G., Willem, M., Hartung, B., Condron, M. M., Teplow, D. B., Haass, C. and Walter, J. (2002). A non-amyloidogenic function of BACE-2 in the secretory pathway. *J Neurochem* 81, 1011-20.

Fluhrer, R., Multhaup, G., Schlicksupp, A., Okochi, M., Takeda, M., Lammich, S., Willem, M., Westmeyer, G., Bode, W., Walter, J. et al. (2003). Identification of a beta-secretase activity, which truncates amyloid beta-peptide after its presenilin-dependent generation. *J Biol Chem* 278, 5531-8.

Goldsbury, C., Mocanu, M. M., Thies, E., Kaether, C., Haass, C., Keller, P., Biernat, J., Mandelkow, E. and Mandelkow, E. M. (2006). Inhibition of APP trafficking by tau protein does not increase the generation of amyloid-beta peptides. *Traffic* 7, 873-88.

Haass, C., Lemere, C. A., Capell, A., Citron, M., Seubert, P., Schenk, D., Lannfelt, L. and Selkoe, D. J. (1995). The Swedish mutation causes early-onset Alzheimer's disease by beta-secretase cleavage within the secretory pathway. *Nat Med* 1, 1291-6.

Haniu, M., Denis, P., Young, Y., Mendiaz, E. A., Fuller, J., Hui, J. O., Bennett, B. D., Kahn, S., Ross, S., Burgess, T. et al. (2000). Characterization of Alzheimer's beta -secretase protein BACE. A pepsin family member with unusual properties. *J Biol Chem* 275, 21099-106.

He, X., Chang, W. P., Koelsch, G. and Tang, J. (2002). Memapsin 2 (beta-secretase) cytosolic domain binds to the VHS domains of GGA1 and GGA2: implications on the endocytosis mechanism of memapsin 2. *FEBS Lett* 524, 183-7.

He, W., Lu, Y., Qahwash, I., Hu, X.Y., Chang, A. And Yan, R. (2004). Reticulon family members modulate BACE1 activity and amyloid-beta peptide generation. *Nat Med* 10, 959-65.

He, X., Li, F., Chang, W. P. and Tang, J. (2005). GGA proteins mediate the recycling pathway of memapsin 2 (BACE). *J Biol Chem* 280, 11696-703.

Holsinger, R. M., McLean, C. A., Beyreuther, K., Masters, C. L. and Evin, G. (2002). Increased expression of the amyloid precursor beta-secretase in Alzheimer's disease. *Ann Neurol* 51, 783-6.

Hu, X., Hicks, C. W., He, W., Wong, P., Macklin, W. B., Trapp, B. D. and Yan, R. (2006). Bace1 modulates myelination in the central and peripheral nervous system. *Nat Neurosci* 9, 1520-5.

Huse, J. T., Pijak, D. S., Leslie, G. J., Lee, V. M. and Doms, R. W. (2000). Maturation and endosomal targeting of beta-site amyloid precursor protein-cleaving enzyme. The Alzheimer's disease beta-secretase. *J Biol Chem* 275, 33729-37.

Hussain, I., Powell, D., Howlett, D. R., Tew, D. G., Meek, T. D., Chapman, C., Gloger, I. S., Murphy, K. E., Southan, C. D., Ryan, D. M. et al. (1999). Identification of a novel aspartic protease (Asp 2) as beta-secretase. *Mol Cell Neurosci* 14, 419-27.

Jacobsen, L., Madsen, P., Nielsen, M. S., Geraerts, W. P., Gliemann, J., Smit, A. B. and Petersen, C. M. (2002). The sorLA cytoplasmic domain interacts with GGA1 and -2 and defines minimum requirements for GGA binding. *FEBS Lett* 511, 155-8.

Kamal, A., Almenar-Queralt, A., LeBlanc, J. F., Roberts, E. A. and Goldstein, L. S. (2001). Kinesin-mediated axonal transport of a membrane compartment containing beta-secretase and presenilin-1 requires APP. *Nature* 414, 643-8.

Kamal, A., Stokin, G. B., Yang, Z., Xia, C. H. and Goldstein, L. S. (2000). Axonal transport of amyloid precursor protein is mediated by direct binding to the kinesin light chain subunit of kinesin-I. *Neuron* 28, 449-59.

Kametaka, S., Shibata, M., Moroe, K., Kanamori, S., Ohsawa, Y., Waguri, S., Sims, P. J., Emoto, K., Umeda, M. and Uchiyama, Y. (2003). Identification of phospholipid scramblase 1 as a novel interacting molecule with beta -secretase (beta -site amyloid precursor protein (APP) cleaving enzyme (BACE)). *J Biol Chem* 278, 15239-45.

Kitazume, S., Tachida, Y., Oka, R., Kotani, N., Ogawa, K., Suzuki, M., Dohmae, N., Takio, K., Saido, T. C. and Hashimoto, Y. (2003). Characterization of alpha 2,6-sialyltransferase cleavage by Alzheimer's beta -secretase (BACE1). *J Biol Chem* 278, 14865-71.

Koh, Y. H., von Arnim, C. A., Hyman, B. T., Tanzi, R. E. and Tesco, G. (2005). BACE is degraded via the lysosomal pathway. *J Biol Chem* 280, 32499-504.

Koo, E. H., Squazzo, S. L., Selkoe, D. J. and Koo, C. H. (1996). Trafficking of cell-surface amyloid beta-protein precursor. I. Secretion, endocytosis and recycling as detected by labeled monoclonal antibody. *J Cell Sci* 109 (Pt 5), 991-8.

Kuhn, P. H., Marjaux, E., Imhof, A., De Strooper, B., Haass, C. and Lichtenthaler, S. F. (2007). Regulated Intramembrane Proteolysis of the Interleukin-1 Receptor II by {alpha}-, beta-, and {gamma}-Secretase. *J Biol Chem* 282, 11982-95.

Laird, F. M., Cai, H., Savonenko, A. V., Farah, M. H., He, K., Melnikova, T., Wen, H., Chiang, H. C., Xu, G., Koliatsos, V. E. et al. (2005). BACE1, a major determinant of selective vulnerability of the brain to amyloid-beta amyloidogenesis, is essential for cognitive, emotional, and synaptic functions. *J Neurosci* 25, 11693-709.

Lazarov, O., Morfini, G. A., Lee, E. B., Farah, M. H., Szodorai, A., DeBoer, S. R., Koliatsos, V. E., Kins, S., Lee, V. M., Wong, P. C. et al. (2005). Axonal transport, amyloid precursor protein, kinesin-1, and the processing apparatus: revisited. *J Neurosci* 25, 2386-95.

Lee, E. B., Zhang, B., Liu, K., Greenbaum, E. A., Doms, R. W., Trojanowski, J. Q. and Lee, V. M. (2005). BACE overexpression alters the subcellular processing of APP and inhibits Abeta deposition in vivo. *J Cell Biol* 168, 291-302.

Li, Q. and Sudhof, T. C. (2004). Cleavage of amyloid-beta precursor protein and amyloid-beta precursor-like protein by BACE 1. *J Biol Chem* 279, 10542-50.

Lichtenthaler, S. F., Dominguez, D. I., Westmeyer, G. G., Reiss, K., Haass, C., Saftig, P., De Strooper, B. and Seed, B. (2003). The cell adhesion protein P-selectin glycoprotein ligand-1 is a substrate for the aspartyl protease BACE1. *J Biol Chem* 278, 48713-9.

Liu, K., Doms, R. W. and Lee, V. M. (2002). Glu11 site cleavage and N-terminally truncated A beta production upon BACE overexpression. *Biochemistry* 41, 3128-36.

Lo, A. C., Haass, C., Wagner, S. L., Teplow, D. B. and Sisodia, S. S. (1994). Metabolism of the "Swedish" amyloid precursor protein variant in Madin-Darby canine kidney cells. *J Biol Chem* 269, 30966-73.

Luo, Y., Bolon, B., Kahn, S., Bennett, B. D., Babu-Khan, S., Denis, P., Fan, W., Kha, H., Zhang, J., Gong, Y. et al. (2001). Mice deficient in BACE1, the Alzheimer's beta-secretase, have normal phenotype and abolished beta-amyloid generation. *Nat Neurosci* 4, 231-2.

Mathews, P. M., Guerra, C. B., Jiang, Y., Grbovic, O. M., Kao, B. H., Schmidt, S. D., Dinakar, R., Mercken, M., Hille-Rehfeld, A., Rohrer, J. et al. (2002). Alzheimer's disease-related overexpression of the cation-dependent mannose 6-phosphate receptor increases Abeta secretion: role for altered lysosomal hydrolase distribution in beta-amyloidogenesis. *J Biol Chem* 277, 5299-307.

Offe, K., Dodson, S. E., Shoemaker, J. T., Fritz, J. J., Gearing, M., Levey, A. I. and Lah, J. J. (2006). The lipoprotein receptor LR11 regulates amyloid beta production and amyloid precursor protein traffic in endosomal compartments. *J Neurosci* 26, 1596-603.

Pastorino, L., Ikin, A. F., Nairn, A. C., Pursnani, A. and Buxbaum, J. D. (2002). The carboxyl-terminus of BACE contains a sorting signal that regulates BACE trafficking but not the formation of total A(beta). *Mol Cell Neurosci* 19, 175-85.

Perez, R. G., Squazzo, S. L. and Koo, E. H. (1996). Enhanced release of amyloid beta-protein from codon 670/671 "Swedish" mutant beta-amyloid precursor protein occurs in both secretory and endocytic pathways. *J Biol Chem* 271, 9100-7.

Puertollano, R., Aguilar, R. C., Gorshkova, I., Crouch, R. J. and Bonifacino, J. S. (2001). Sorting of mannose 6-phosphate receptors mediated by the GGAs. *Science* 292, 1712-6.

Qing, H., Zhou, W., Christensen, M. A., Sun, X., Tong, Y. and Song, W. (2004). Degradation of BACE by the ubiquitin-proteasome pathway. *Faseb J* 18, 1571-3.

Rentmeister, A., Bill, A., Wahle, T., Walter, J. and Famulok, M. (2006). RNA aptamers selectively modulate protein recruitment to the cytoplasmic domain of beta-secretase BACE1 in vitro. *Rna* 12, 1650-60.

Roberds, S. L., Anderson, J., Basi, G., Bienkowski, M. J., Branstetter, D. G., Chen, K. S., Freedman, S. B., Frigon, N. L., Games, D., Hu, K. et al. (2001). BACE knockout mice are healthy despite lacking the primary beta-secretase activity in brain: implications for Alzheimer's disease therapeutics. *Hum Mol Genet* 10, 1317-24.

Rockenstein, E., Mante, M., Alford, M., Adame, A., Crews, L., Hashimoto, M., Esposito, L., Mucke, L. and Masliah, E. (2005). High beta-secretase activity elicits neurodegeneration in transgenic mice despite reductions in amyloid-beta levels: implications for the treatment of Alzheimer disease. *J Biol Chem* 280, 32957-67.

Rogaeva, E., Meng, Y., Lee, J. H., Gu, Y., Kawarai, T., Zou, F., Katayama, T., Baldwin, C. T., Cheng, R., Hasegawa, H. et al. (2007). The neuronal sortilin-related receptor SORL1 is genetically associated with Alzheimer disease. *Nat Genet* 39, 168-177.

Saunders, A. M., Schmader, K., Breitner, J. C., Benson, M. D., Brown, W. T., Goldfarb, L., Goldgaber, D., Manwaring, M. G., Szymanski, M. H., McCown, N. et al. (1993). Apolipoprotein E epsilon 4 allele distributions in late-onset Alzheimer's disease and in other amyloid-forming diseases. *Lancet* 342, 710-1.

Schmechel, A., Strauss, M., Schlicksupp, A., Pipkorn, R., Haass, C., Bayer, T. A. and Multhaup, G. (2004). Human BACE forms dimers and colocalizes with APP. *J Biol Chem* 279, 39710-7.

Selkoe, D. J. (2001). Alzheimer's disease: genes, proteins, and therapy. *Physiol Rev* 81, 741-66.

Shi, X. P., Chen, E., Yin, K. C., Na, S., Garsky, V. M., Lai, M. T., Li, Y. M., Platchek, M., Register, R. B., Sardana, M. K. et al. (2001). The pro domain of beta-secretase does not confer strict zymogen-like properties but does assist proper folding of the protease domain. *J Biol Chem* 276, 10366-73.

Shi, X. P., Tugusheva, K., Bruce, J. E., Lucka, A., Wu, G. X., Chen-Dodson, E., Price, E., Li, Y., Xu, M., Huang, Q. et al. (2003). Beta-secretase cleavage at amino acid residue 34 in the amyloid beta peptide is dependent upon gamma-secretase activity. *J Biol Chem* 278, 21286-94.

Simons, K. and Toomre, D. (2000). Lipid rafts and signal transduction. Nat Rev *Mol Cell Biol* 1, 31-9.

Sinha, S., Anderson, J. P., Barbour, R., Basi, G. S., Caccavello, R., Davis, D., Doan, M., Dovey, H. F., Frigon, N., Hong, J. et al. (1999). Purification and cloning of amyloid precursor protein beta-secretase from human brain. *Nature* 402, 537-40.

Spoelgen, R., von Arnim, C. A., Thomas, A. V., Peltan, I. D., Koker, M., Deng, A., Irizarry, M. C., Andersen, O. M., Willnow, T. E. and Hyman, B. T. (2006). Interaction of the cytosolic domains of sorLA/LR11 with the amyloid precursor protein (APP) and beta-secretase beta-site APP-cleaving enzyme. *J Neurosci* 26, 418-28.

Steiner, H. and Haass, C. (2000). Intramembrane proteolysis by presenilins. *Nat Rev Mol Cell Biol* 1, 217-24.

Thinakaran, G., Teplow, D. B., Siman, R., Greenberg, B. and Sisodia, S. S. (1996). Metabolism of the "Swedish" amyloid precursor protein variant in neuro2a (N2a) cells. Evidence that cleavage at the "beta-secretase" site occurs in the golgi apparatus. *J Biol Chem* 271, 9390-7.

Valentine, J. S. and Hart, P. J. (2003). Misfolded CuZnSOD and amyotrophic lateral sclerosis. *Proc Natl Acad Sci U S A* 100, 3617-22.

Vassar, R. (2001). The beta-secretase, BACE: a prime drug target for Alzheimer's disease. *J Mol Neurosci* 17, 157-70.

Vassar, R., Bennett, B. D., Babu-Khan, S., Kahn, S., Mendiaz, E. A., Denis, P., Teplow, D. B., Ross, S., Amarante, P., Loeloff, R. et al. (1999). Beta-secretase cleavage of Alzheimer's amyloid precursor protein by the transmembrane aspartic protease BACE. *Science* 286, 735-41.

von Arnim, C. A., Kinoshita, A., Peltan, I. D., Tangredi, M. M., Herl, L., Lee, B. M., Spoelgen, R., Hshieh, T. T., Ranganathan, S., Battey, F. D. et al. (2005). The low density lipoprotein receptor-related protein (LRP) is a novel beta-secretase (BACE1) substrate. *J Biol Chem* 280, 17777-85.

von Arnim, C. A., Spoelgen, R., Peltan, I. D., Deng, M., Courchesne, S., Koker, M., Matsui, T., Kowa, H., Lichtenthaler, S. F., Irizarry, M. C. et al. (2006). GGA1 acts as a spatial switch altering amyloid precursor protein trafficking and processing. *J Neurosci* 26, 9913-22.

von Arnim, C. A., Tangredi, M. M., Peltan, I. D., Lee, B. M., Irizarry, M. C., Kinoshita, A. and Hyman, B. T. (2004). Demonstration of BACE (beta-secretase) phosphorylation and its interaction with GGA1 in cells by fluorescence-lifetime imaging microscopy. *J Cell Sci* 117, 5437-45.

Wahle, T., Prager, K., Raffler, N., Haass, C., Famulok, M. and Walter, J. (2005). GGA proteins regulate retrograde transport of BACE1 from endosomes to the trans-Golgi network. *Mol Cell Neurosci* 29, 453-61.

Wahle, T., Thal, D. R., Sastre, M., Rentmeister, A., Bogdanovic, N., Famulok, M., Heneka, M. T. and Walter, J. (2006). GGA1 is expressed in the human brain and affects the generation of amyloid beta-peptide. *J Neurosci* 26, 12838-46.

Walter, J., Capell, A., Grunberg, J., Pesold, B., Schindzielorz, A., Prior, R., Podlisny, M. B., Fraser, P., Hyslop, P. S., Selkoe, D. J. et al. (1996). The Alzheimer's disease-associated presenilins are differentially phosphorylated proteins located predominantly within the endoplasmic reticulum. *Mol Med* 2, 673-91.

Walter, J., Fluhrer, R., Hartung, B., Willem, M., Kaether, C., Capell, A., Lammich, S., Multhaup, G. and Haass, C. (2001). Phosphorylation regulates intracellular trafficking of beta-secretase. *J Biol Chem* 276, 14634-41.

Walter, J., Kaether, C., Steiner, H. and Haass, C. (2001). The cell biology of Alzheimer's disease: uncovering the secrets of secretases. *Curr Opin Neurobiol* 11, 585-90.

Westmeyer, G. G., Willem, M., Lichtenthaler, S. F., Lurman, G., Multhaup, G., Assfalg-Machleidt, I., Reiss, K., Saftig, P. and Haass, C. (2004). Dimerization of beta-site beta-amyloid precursor protein-cleaving enzyme. *J Biol Chem* 279, 53205-12.

Willem, M., Garratt, A. N., Novak, B., Citron, M., Kaufmann, S., Rittger, A., DeStrooper, B., Saftig, P., Birchmeier, C. and Haass, C. (2006). Control of peripheral nerve myelination by the beta-secretase BACE1. *Science* 314, 664-6.

Wong, H. K., Sakurai, T., Oyama, F., Kaneko, K., Wada, K., Miyazaki, H., Kurosawa, M., De Strooper, B., Saftig, P. and Nukina, N. (2005). beta Subunits of voltage-gated sodium channels are novel substrates of beta-site amyloid precursor protein-cleaving enzyme (BACE1) and gamma-secretase. *J Biol Chem* 280, 23009-17.

Wong, P. C., Waggoner, D., Subramaniam, J. R., Tessarollo, L., Bartnikas, T. B., Culotta, V. C., Price, D. L., Rothstein, J. and Gitlin, J. D. (2000). Copper chaperone for superoxide dismutase is essential to activate mammalian Cu/Zn superoxide dismutase. *Proc Natl Acad Sci U S A* 97, 2886-91.

Yamazaki, T., Koo, E. H. and Selkoe, D. J. (1996). Trafficking of cell-surface amyloid beta-protein precursor. II. Endocytosis, recycling and lysosomal targeting detected by immunolocalization. *J Cell Sci* 109 (Pt 5), 999-1008.

Yan, R., Bienkowski, M. J., Shuck, M. E., Miao, H., Tory, M. C., Pauley, A. M., Brashier, J. R., Stratman, N. C., Mathews, W. R., Buhl, A. E. et al. (1999). Membrane-anchored aspartyl protease with Alzheimer's disease beta-secretase activity. *Nature* 402, 533-7.

Yang, L. B., Lindholm, K., Yan, R., Citron, M., Xia, W., Yang, X. L., Beach, T., Sue, L., Wong, P., Price, D. et al. (2003). Elevated beta-secretase expression and enzymatic activity detected in sporadic Alzheimer disease. *Nat Med* 9, 3-4.

In:Cognitive Science Compendium Volume 2
Editors: Miao-Kun Sun

ISBN: 978-1-60456-775-5

Chapter 2

BRAIN VESICULAR MONOAMINE TRANSPORTER AND APOPTOSIS: PERSPECTIVES ON DEVELOPMENT AND NEURODEGENERATION

Léa Stankovski[1], Patricia Gaspar[1] and Olivier Cases[1,2,*]

[1.] INSERM, UMR616, Université Pierre et Marie Curie, Hôpital de la Pitié-Salpêtrière, 47 Boulevard de l'Hôpital, 75013 Paris, France.

[2.] INSERM, UMR676, Université Denis Diderot, Hôpital Robert Debré, 48 Boulevard Sérurier, 75019 Paris, France.

ABSTRACT

The neuronal isoform of vesicular monoamine transporter, VMAT2, is responsible for packaging dopamine, norepinephrine, and serotonin into synaptic vesicles and thereby plays an essential role in monoamine neurotransmission. This sequestering action is important for normal synaptic release of monoamines but it may also act to keep intracellular levels of the monoamine transmitters below potentially toxic levels. During embryonic and postnatal development, VMAT2 is expressed in many non-aminergic neurons in the brain, long before synapses are formed, suggesting that it could have non-synaptic roles. Here, we review recent evidences indicating a role for VMAT2 in the control of developmental cell death in the cerebral cortex and in models of Parkinson related disorders. Abnormalities in VMAT2 functions have been suggested to play a key role in the etiology of a number of disorders, including Parkinson's disease and addiction.

Keywords: vesicular monoamine transporter 2, monoamines, apoptosis, cerebral cortex, Parkinson's disease.

---

* Correspondence and requests for reprints: Dr Olivier Cases, UMR676, Hôpital Robert Debré, 49 Boulevard Sérurier, 75019 Paris, France. Email: olivier.cases@rdebre.inserm.fr

INTRODUCTION

Vesicular neurotransmitter transporters are essential components of secretory vesicle, which are responsible for storage and regulated secretion of neurotransmitters in neurons and neuroendocrine cells. The nature of the vesicular transporters expressed by a given cell dictates what kind, and how much neurotransmitter is released during synaptic transmission (Eiden, 2000). Vesicular transporters are members of the toxin-extruding proton-translocating antiporter family. As such, another possibly important function of these transporters is to protect the cells from accumulating potentially toxic levels of transmitters intracellularly.

The vesicular monoamine transporters (VMAT) are responsible for the translocation of biogenic monoamines (serotonin, dopamine, norepinephrine, and histamine) from the cytoplasm into synaptic vesicles via a proton electrochemical gradient generated by the vacuolar type H+-adenosine triphosphate. Two vesicular monoamine transporters have been identified by expression, cloning, and pharmacology (Erickson et al., 1992; Liu et al., 1992; Erickson and Eiden, 1993). VMAT1 and VMAT2 arise from two separate genes, yet have extensive homology.

Only VMAT1 is expressed in serotonin-accumulating enterochromaffin cells, whereas only VMAT2 is found in enterochromaffin-like histaminocytes of the stomach (Eiden, 2000). Differences in the substrate recognition and inhibitor sensitivities between VMAT1 and VMAT2 have been studied, using membrane vesicles prepared from stable transformed cell lines from Chinese hamster ovaries (CHO) that express the respective proteins (Peter et al., 1994). The major difference between the two transporters is that VMAT2 efficiently transports histamine and VMAT1 does not. VMAT1 may have evolved during the evolutionary emergence of histamine as a secreted effector molecule, restricting the storage of histamine in enterochromaffin cells of the lower gut (Hoffman et al., 1998).

Both VMAT1 and VMAT2 are more widely expressed in the brain during embryonic and early postnatal development .

VMAT2 VERSUS MONOAMINE OXIDASE

Oxidative deamination of monoamines by monoamine oxidase type A (MAOA) and type B (MAOB) is accompanied by the reduction of molecular oxygen to a toxic product, hydrogen peroxide (Shih et al., 1999). Therefore, maintenance of low cytoplasmic concentrations of neurotransmitters by their reuptake into synaptic vesicles is important to minimize their inherent toxicity. Furthermore, storage of neurotransmitters in synaptic vesicles precludes their metabolism in the cytoplasmic compartment and reduces the synthetic demands on the cell. In the central nervous system, VMAT2 is the only transporter that moves cytoplasmic monoamines into synaptic vesicles for storage and subsequent exocytotic release. MAOA has higher affinity for the substrates serotonin (5-HT), norepinephrine (NE), dopamine (DA), histamine (HA), and the inhibitor clorgyline, whereas MAO B has higher affinity for phenylethylamine (PEA), benzylamine, and the inhibitor deprenyl.

MAOA and MAOB display non-overlapping patterns of mRNA and protein expression. MAOA is abundantly expressed in noradrenergic, and dopaminergic neurons and restricted neuronal populations. Conversely, MAOB expression is highly expressed in serotoninergic and histaminergic neurons, as well as in astrocytes (Vitalis et al., 2002).

DEVELOPMENTAL EXPRESSION OF VMAT2

In all the norepinephrine-, epinephrine-, serotonin-, dopamine-, and histamine-containing cell groups, VMAT2 is expressed one day after embryonic differentiation; i.e., E13-E14 for serotoninergic neurons of the rat dorsal raphe (Hansson et al., 1998a,b). In addition, VMAT2 is expressed in selective non-monoaminergic neuronal populations during embryonic and postnatal development. These neurons are not enzymatically equipped to produce amines and instead appear to have a glutamatergic phenotype. VMAT2 is expressed during a defined period in each neuronal population. For instance, the glutamatergic neurons of the sensory thalamus express VMAT2 from E15 to P15, whereas in the cerebral cortex, VMAT2 is expressed from E18 to P5 (Hansson et al., 1998a,b, Lebrand et al., 1998). In most neuronal populations, VMAT2 is co-expressed with the plasma membrane serotonin transporter (Lebrand et al., 1998), suggesting that it is capable of handling serotonin. Curiously, a few populations expressing VMAT2 do not display other known monoaminergic markers. However, all the VMAT2-expressing neurons are capable of accumulating serotonin, under the conditions where the extracellular levels of this amine are raised (Cases et al., 1998).

MICE LACKING VMAT2

Genetic models, such as those afforded by the VMAT2 deficient mice (Fon et al., 1997; Takahashi et al., 1997; Wang et al., 1997) allow the re-evaluation of the effects of VMAT2 depletion without the inherent limitations of pharmacological treatments. Homozygous VMAT2 KO mice are severely atrophic and most pups die by P4 (Fon et al., 1997; Takahashi et al., 1997; Wang et al., 1997). VMAT2 KO pups are born in the expected Mendelian ratio and their gross brain morphology appears normal. In addition, monoamine cell groups in the brainstem and their projections have a normal appearance. In particular, DA islands and the patch-matrix system appear unaffected by the elimination of VMAT2, suggesting that monoamine release does not have a role in the formation of these connections (Fon et al., 1997). Nonetheless, VMAT2 KO pups move little and feed poorly compared to their wild-type and heterozygous littermates, accounting for their growth failure and early postnatal mortality. Consistent with the role for VMAT2, DA neurons cultured from the VMAT2 KO as well as striatal slices prepared from these animals show no depolarization-evoked DA release (Fon et al., 1997; Wang et al., 1997). However, amphetamine still induces DA release from VMAT2 KO neurons, indicating that VMAT2 is not required for the action of the psychostimulant, at least under these circumstances where cytoplasmic DA may already be elevated. Indeed, administration of amphetamine to VMAT2 KO increases movements, such as walking and feeding, and enables the animals to survive for several weeks (Fon et al., 1997). Thus, precisely regulated exocytotic release of monoamines does not seem to be required for several complex behaviours.

Disruption of the VMAT2 gene results in dramatically reduced brain levels of DA, NE, and 5-HT (Fon et al., 1997; Wang et al., 1997). This could be a consequence of reduced monoamine synthesis or increased degradation. However, disruption of VMAT2 does not alter the expression level of the rate-limiting catecholamine biosynthetic enzyme tyrosine hydroxylase and TH activity appears to be upregulated (Wang et al., 1997). Furthermore, monoamine metabolites are not reduced in the VMAT2 KO mice (Alvarez et al., 2002). All these observations suggest that VMAT2 ablation causes an increase in monoamine catabolism rather that a decrease in synthesis. Interestingly, an administration of the MAOA inhibitor clorgyline increases 5-HT levels, supporting this interpretation. However, clorgyline has no effect on DA levels, indicating that additional mechanisms act to control the level of this neurotransmitter (Fon et al., 1997). In VMAT2-MAOA DKO mouse, pups are severely hypomorphic in comparison to the wild-type or MAOA KO mice but they survive until P13 (Alvarez et al., 2002).

VMAT2 and Programmed Cell Death in the Cerebral Cortex

The development of the cerebral cortex is a sequential process that includes programmed cell death (PCD). Approximately half of the neurons produced during corticogenesis are thought to die (Ferrer et al., 1992; Blaschke et al., 1996). Two consecutive waves of PCD affect the cortical neurons at different periods of their development. The first wave consists of cell death of the proliferating precursors in the ventricular and subventricular zones, which appears to be closely linked to cell cycle regulation (Thomaidou et al., 1997). The second wave affects postmitotic neurons at later stages and may be involved in matching the size of neuronal populations to that of their targets during the formation and the maintenance of synapses (Ferrer et al., 1992). At both periods, cell death is apoptotic and the molecular machinery relies on the mitochondrial pathways of intracellular signal transduction (Dikranian et al., 2001).

Using this model, we found a significant increase of cell death in the supragranular layers of the cingulate and retrosplenial areas (Skankovski et al., 2007). This effect is regionally specific since some regions such as the primary somatosensory cortex show no changes (Alvarez et al. 2002), whereas other regions, such as the hippocampus or striatum, display an increase in cell death (unpublished). Electron microscopic analysis and TUNEL staining confirmed that this was related to an increase in apoptosis. Additional characterization indicated an increase in caspase-9-, and caspase-3-immunoreactive profiles, suggesting that the increase in cell death observed in the cortex of VMAT2 KO mice is attributable to an increase of apoptotic caspase-dependent mechanism. We also observed a specific decrease in the mRNA levels of the anti-apoptotic factor Bcl-XL, whereas mRNA levels of the proapoptotic factor Bax remained unchanged. This suggests that the decrease of the Bcl-XL/Bax mRNA ratio is probably the main phenomenon at play in the increased cell death in VMAT2 KO cerebral cortex (Stankovski et al., 2007). However, the precise mechanisms by which the depletion of VMAT2 leads to this change in gene expression and to exaggerated cell death remain elusive at the moment.

In synaptic vesicles, monoamines are protected from degradation by MAO activity. MAOA is the major form expressed in specific neuronal populations, such as the serotonergic neurons during early postnatal life (Vitalis et al., 2002). Thus, invalidating MAOA in a VMAT2 KO genetic background should maintain a pool of cytoplasmic 5-HT by preventing its degradation. Indeed, high serotonin levels were detected in the brain of VMAT2-MAOA DKO mice, whereas dopamine and norepinephrine levels remained low or undetectable (Alvarez et al., 2002). Interestingly, the increase in cell death observed in the VMAT2 KO was reversed in the cerebral cortex of VMAT2-MAOA DKO, indicating that increases in brain serotonin levels could have a neuroprotectant effect on cortical neurons (Stankovski et al., 2007). This effect is in agreement with a previous report by Persico et al. (2003) in the brain of mice lacking 5-HTT. 5-HTT KO mice display high serotonin levels and reduced PCD, as judged by the decrease of TUNEL-positive cells in their brain. We found that Bcl-XL mRNA levels returned to normal in the cerebral cortex of VMAT2-MAOA DKO, and no decrease in expression of the proapoptotic factor Bax was noted (Stankovski et al., 2007). This suggests that high serotonin levels maintain a normal Bcl-XL/Bax mRNA ratio. As a consequence, a normal level of cell death is observed in the cerebral cortex of VMAT2-MAOA DKO. The mechanisms that allow this normalization at both the cellular and molecular levels are not fully understood but likely involve one or several 5-HT receptor subtypes. We found that 5-HT2 receptor activation led to a partial restoration of the increased cell death phenotype in VMAT2 KO (Stankovski et al., 2007). These results are consistent with the prosurvival activity of 5-HT2 receptor activation that was noted previously in vitro (Dooley et al., 1997). Because activation of 5-HT2A/2C receptors leads to an augmentation of the trophic factor, brain-derived neurotrophic factor (BDNF) (Vaidya et al., 1997) and mice lacking the catalytic domain of trkB, the high-affinity receptor of BDNF, display an increase in cell death in the cingulate cortex (Alcantara et al., 1997), we hypothesized that trkB-dependent mechanisms could be involved in the anti-apoptotic cascade triggered by high serotonin levels. However, quantification of BDNF mRNA levels showed no differences between normal, VMAT2 KO, or VMAT2-MAOA DKO in the cingulate cortex. Furthermore, trkB KO showed a rescue of cell death in conditions of increased 5-HT levels (Stankovski et al., 2007). These results suggest that cell death in the VMAT2 KO is not attributable to a decrease of trkB signalling. However, we cannot rule out signalling via p75 neurotrophin receptor (Kalb, 2005).

Parkinson's Disease

Parkinson's disease is a progressively degenerative disorder that affects preferentially neurons of the substantia nigra. In this regard, DA may play a role as an endogenous toxin, since the normal metabolism of DA produces hydrogen peroxide as a by-product (Adams et al., 2001). Accordingly, pharmacological enhancement of DA sequestration by VMAT2, to prevent oxidation of DA in the cytoplasm, could be an interesting strategy for treatment of Parkinson's disease.

Exposure to the neurotoxin N-methyl-4-phenyltetrahydropyridine (MPTP) results in clinical symptoms that resemble Parkinson's disease (Langston, 1996). N-Methyl-4-phenylpyridinium (MPP+), the active toxic metabolite of MPTP, is a substrate for VMAT2

(Scherman et al., 1988; Moriyama et al., 1993). VMAT2 sequesters MPP+ in synaptic vesicles and thereby protects catecholamine-containing neurons from MPP+-induced toxicity and degeneration (Takahashi et al., 1997; Speciale et al., 1998; German et al., 2000; Staal et al., 2000). Studies using heterozygous VMAT2 knockout mice show that the knockouts are more susceptible to the neurotoxic effects of MPTP, compared with the wild-type mice (Takahashi et al., 1997; Gainetdinov et al., 1998; Mooslehner et al., 2001). Furthermore, heterozygous VMAT2 knockout mice are more sensitive to methamphetamine-induced neurotoxicity and are more vulnerable to the toxic effects of L-3,4-dihydroxyphenylalanine (L-DOPA, a DA precursor used to treat Parkinson's disease), compared with wild-type mice (Fumagalli et al., 1999; Kariya et al., 2005). The latter results suggest that reduction in VMAT2 activity might attenuate the efficacy of L-DOPA therapy in Parkinson's patients. Finally, increased sequestration of DA in synaptic vesicles by VMAT2 has been suggested to be protective in Parkinson's disease (Glatt et al., 2006).

Taken together, the results of the above studies indicate that VMAT2 expression and function are important in counteracting the neurotoxicity of MPP+ and perhaps of other environmental and endogenous neurotoxins that play an etiologic role in neurodegenerative diseases.

Sponsors

This work was supported by Institut National de la Santé et de la Recherche Médicale, Centre National de la Recherche Scientifique (O.C.), University Paris 6, and Agence Nationale de la Recherche (-05- APV05146DSA).

References

Alcantara S, Frisen J, del Rio JA, Soriano E, Barbacid M, Silos-Santiago I (1997) TrkB signaling is required for postnatal survival of CNS neurons and protects hippocampal and motor neurons from axotomy-induced cell death. *J Neurosci* 17:3623-3633.

Alvarez C, Vitalis T, Fon EA, Hanoun N, Hamon M, Seif I, Edwards R, Gaspar P, Cases O (2002) Effects of genetic depletion of monoamines on somatosensory cortical development. *Neuroscience* 115:753-764.

Blaschke AJ, Staley K, Chun J (1996) Widespread programmed cell death in proliferative and postmitotic regions of the fetal cerebral cortex. *Development* 122:1165-1174.

Cases O, Lebrand C, Giros B, Vitalis T, De Maeyer E, Caron MG, Price DJ, Gaspar P, Seif I (1998) Plasma membrane transporters of serotonin, dopamine, and norepinephrine mediate serotonin accumulation in atypical locations in the developing brain of monoamine oxidase A knock-outs. *J Neurosci* 18: 6914-6927.

Dikranian K, Ishimaru MJ, Tenkova T, Labruyere J, Qin YQ, Ikonomidou C, Olney JW (2001) Apoptosis in the in vivo mammalian forebrain. *Neurobiol Dis* 8:359-379.

Dooley AE, Pappas IS, Parnavelas JG (1997) Serotonin promotes the survival of cortical glutamatergic neurons in vitro. *Exp Neurol* 148:205-214.

Erickson JD, Eiden LE (1993) Functional identification and molecular cloning of a human brain vesicle monoamine transporter. *J Neurochem* 61:2314-2317.

Erickson JD, Eiden LE, Hoffman BJ (1992) Expression cloning of a reserpine-sensitive vesicular monoamine transporter. *Proc Natl Acad Sci U S A* 89:10993-10997.

Ferrer I, Soriano E, del Rio JA, Alcantara S, Auladell C (1992) Cell death and removal in the cerebral cortex during development. *Prog Neurobiol* 39:1-43.

Fon EA, Pothos EN, Sun BC, Killeen N, Sulzer D, Edwards RH (1997) Vesicular transport regulates monoamine storage and release but is not essential for amphetamine action. *Neuron* 19:1271-1283.

Fumagalli F, Gainetdinov RR, Wang YM, Valenzano KJ, Miller GW, Caron MG (1999) Increased methamphetamine neurotoxicity in heterozygous vesicular monoamine transporter 2 knock-out mice. *J Neurosci* 19:2424-2431.

Gainetdinov RR, Fumagalli F, Wang YM, Jones SR, Levey AI, Miller GW, Caron MG (1998) Increased MPTP neurotoxicity in vesicular monoamine transporter 2 heterozygote knockout mice. *J Neurochem* 70:1973-1978.

German DC, Liang CL, Manaye KF, Lane K, Sonsalla PK (2000) Pharmacological inactivation of the vesicular monoamine transporter can enhance 1-methyl-4-phenyl-1,2,3,6-tetrahydropyridine-induced neurodegen-eration of midbrain dopaminergic neurons, but not locus coeruleus noradrenergic neurons. *Neuroscience* 101:1063-1069.

Glatt CE, Wahner AD, White DJ, Ruiz-Linares A, Ritz B (2006) Gain-of-function haplotypes in the vesicular monoamine transporter promoter are protective for Parkinson disease in women. *Hum Mol Genet* 15:299-305.

Hansson SR, Hoffman BJ, Mezey E (1998a) Ontogeny of vesicular monoamine transporter mRNAs VMAT1 and VMAT2. I The developing rat central nervous system. *Dev Brain Res* 110: 135-158.

Hansson SR, Mezey E, Hoffman BJ (1998b) Ontogeny of vesicular monoamine transporter mRNAs VMAT1 and VMAT2. II Expression in neural crest derivatives and their target sites in the rat. *Dev Brain Res* 110: 159-174.

Hoffman BJ, Hansson SR, Mezey E, Palkovits M (1998) Localization and dynamic regulation of biogenic amine transporters in the mammalian central nervous system. *Front Neuroendocrinol* 19:187-231.

Kalb R (2005) The protean actions of neurotrophins and their receptors on the life and death of neurons. *Trends Neurosci* 28:5-11.

Kariya S, Takahashi N, Hirano M, Ueno S (2005) Increased vulnerability to L-DOPA toxicity in dopaminergic neurons From VMAT2 heterozygote knockout mice. *J Mol Neurosci* 27:277-279.

Langston JW (1996) The etiology of Parkinson's disease with emphasis on the MPTP story. *Neurology* 47:S153-160.

Lebrand C, Cases O, Wehrle R, Blakely RD, Edwards RH, Gaspar P (1998) Transient developmental expression of monoamine transporters in the rodent forebrain. *J Comp Neurol* 401:506-524.

Liu Y, Peter D, Roghani A, Schuldiner S, Prive GG, Eisenberg D, Brecha N, Edwards RH (1992) A cDNA that suppresses MPP+ toxicity encodes a vesicular amine transporter. *Cell* 70:539-551.

Mooslehner KA, Chan PM, Xu W, Liu L, Smadja C, Humby T, Allen ND, Wilkinson LS, Emson PC (2001) Mice with very low expression of the vesicular monoamine transporter

2 gene survive into adulthood: potential mouse model for parkinsonism. *Mol Cell Biol* 21:5321-5331.

Moriyama Y, Amakatsu K, Futai M (1993) Uptake of the neurotoxin, 4-methylphenylpyridinium, into chromaffin granules and synaptic vesicles: a proton gradient drives its uptake through monoamine transporter. *Arch Biochem Biophys* 305:271-277.

Persico AM, Baldi A, Dell'Acqua ML, Moessner R, Murphy DL, Lesch KP, Keller F (2003) Reduced programmed cell death in brains of serotonin transporter knockout mice. *Neuroreport* 14:341-344.

Peter D, Jimenez J, Liu Y, Kim J, Edwards RH (1994) The chromaffin granule and synaptic vesicle amine transporters differ in substrate recognition and sensitivity to inhibitors. *J Biol Chem* 269:7231-7237.

Scherman D, Darchen F, Desnos C, Henry JP (1988) 1-Methyl-4-phenylpyridinium is a substrate of the vesicular monoamine uptake system of chromaffin granules. *Eur J Pharmacol* 146:359-360.

Shih JC, Chen K, Ridd MJ (1999) Monoamine oxidase: from genes to behavior. *Annu Rev Neurosci* 22:197-217.

Speciale SG, Liang CL, Sonsalla PK, Edwards RH, German DC (1998) The neurotoxin 1-methyl-4-phenylpyridinium is sequestered within neurons that contain the vesicular monoamine transporter. *Neuroscience* 84:1177-1185.

Staal RG, Hogan KA, Liang CL, German DC, Sonsalla PK (2000) In vitro studies of striatal vesicles containing the vesicular monoamine transporter (VMAT2): rat versus mouse differences in sequestration of 1-methyl-4-phenylpyridinium. *J Pharmacol Exp Ther* 293:329-335.

Stankovski L, Alvarez C, Ouimet T, Vitalis T, El-Hachimi KH, Price D, Deneris E, Gaspar P, Cases O (2007) Developmental cell death is enhanced in the cerebral cortex of mice lacking the brain vesicular monoamine transporter. *J Neurosci* 27:1315-1324.

Takahashi N, Miner LL, Sora I, Ujike H, Revay RS, Kostic V, Jackson-Lewis V, Przedborski S, Uhl GR (1997) VMAT2 knockout mice: heterozygotes display reduced amphetamine-conditioned reward, enhanced amphetamine locomotion, and enhanced MPTP toxicity. *Proc Natl Acad Sci U S A* 94:9938-9943.

Thomaidou D, Mione MC, Cavanagh JF, Parnavelas JG (1997) Apoptosis and its relation to the cell cycle in the developing cerebral cortex. *J Neurosci* 17:1075-1085.

Vaidya VA, Marek GJ, Aghajanian GK, Duman RS (1997) 5-HT2A receptor-mediated regulation of brain-derived neurotrophic factor mRNA in the hippocampus and the neocortex. *J Neurosci* 17:2785-2795.

Vitalis T, Fouquet C, Alvarez C, Seif I, Price D, Gaspar P, Cases O (2002) Developmental expression of monoamine oxidases A and B in the central and peripheral nervous systems of the mouse. *J Comp Neurol* 442:331-347.

Wang YM, Gainetdinov RR, Fumagalli F, Xu F, Jones SR, Bock CB, Miller GW, Wightman RM, Caron MG (1997) Knockout of the vesicular monoamine transporter 2 gene results in neonatal death and supersensitivity to cocaine and amphetamine. *Neuron* 19:1285-1296.

In:Cognitive Science Compendium Volume 2
Editors: Miao-Kun Sun

ISBN: 978-1-60456-775-5

Chapter 3

Mild Cognitive Impairment is Too Late: The Case for Presymptomatic Detection and Treatment of Alzheimer's Disease

Charles D. Smith
Department of Neurology & Alzheimer's Disease Center, University of Kentucky College of Medicine, Lexington, KY
United States of America

Abstract

The thesis of this review is that the earliest cognitive symptoms in dementia represent the exhaustion of compensatory mechanisms in the brain which counteract underlying Alzheimer's disease (AD), vascular, Lewy body, and other neuropathology. These pathologies may accumulate gradually over years and perhaps enter a phase of acceleration as compensatory resources are outstripped. Evidence for this model of a prolonged presymptomatic period in AD has come from recent imaging, neuropathologic, and basic science studies. These studies and the potential consequences for diagnosis and treatment are presented. The conclusion is that using onset symptoms as the signal to begin disease-modifying treatment for AD is too late; this treatment must begin earlier, before symptoms begin, to preserve brain function. Therefore presymptomatic detection is a critical research goal in AD.

Key Words: Alzheimer's disease, prevention, review, human studies, mild cognitive impairment, neural plasticity

1. Introduction

Presymptomatic detection is different from the concept of detecting cognitive or clinical abnormalities prior to the diagnosis of dementia, or even the very earliest symptoms that presage dementia years later. It refers to detecting the presence of an underlying pathologic process that later expresses itself as a dementia when persons are manifestly normal. The many reasons to suspect this is possible are described here.

This is not an unbiased and dispassionate survey of literature on this topic. It has a specific viewpoint. The thesis is that the earliest cognitive symptoms in dementia represent the exhaustion of compensatory mechanisms in the brain which counteract accumulating Alzheimer's disease (AD), vascular, Lewy body, and other neuropathology. Detecting the presence of an underlying asymptomatic pathology may be indirect, e.g., one may be detecting brain compensations for pathology that would otherwise be unobservable. Though indirect, it is a form of presymptomatic detection.

The motivation for addressing presymptomatic detection now is the imminent availability of disease-modifying treatments for AD. Detecting threshold amounts of AD pathology before symptoms appear offers the hope that symptoms could be delayed or even prevented by applying these treatments in persons with this pathology. This threshold may be determined by several contributions to brain reserve discussed in this article. A summary argument for presymptomatic detection and treatment is presented in the conclusion.

2. The Brain Reserve Hypothesis

The brain reserve hypothesis is the theoretical construct that inherited, developmental and cultural factors influence the symptomatic expression of specific underlying brain pathologies. Examples of such neuropathologies include human immunodeficiency virus (HIV) encephalopathy (Stern, Silva, Chaisson, & Evans, 1996) and neurodegenerative diseases, e.g., Alzheimer's disease (Mortimer, 1997; Mortimer, Borenstein, Gosche, & Snowdon, 2005; Mosconi et al., 2005; Wolf, Julin, Gertz, Winblad, & Wahlund, 2004) and frontotemporal dementia (Perneczky, Diehl-Schmid, Drzezga, & Kurz, 2007; Perneczky, Diehl-Schmid, Pohl, Drzezga, & Kurz, 2007). Examples of factors constituting reserve are years of education (McDowell, Xi, Lindsay, & Tierney, 2007), socioeconomic status level (Stern et al., 1994), head circumference (Borenstein Graves et al., 2001), and intelligence quotient. The neurobiologic basis of each these factors' contribution to reserve is often unstated or heuristically justified in the absence of scientific data other than that from studies demonstrating association between the factor and symptom expression. Nonetheless it is quite plausible that intrinsic or acquired brain properties may confer resistance to the effects of neuropathology, and therefore modify whether and when that neuropathology is expressed as symptoms, changes in cognitive performance or behavior.

2.1. Normal Developmental Changes in the Brain during Childhood and Adolescence

Here the focus is on the key findings related to the concept of brain reserve as reflected in changes in brain anatomy and histology, tissue density, diffusivity, and functional characteristics from conception to early adulthood.

In utero measurements by magnetic resonance imaging (MRI) and ultrasound demonstrate rapid, approximately linear, growth of the brain from 10 to 40 weeks gestation

(Figure 1), corresponding to neurogenesis, and later in the second and third trimesters, increasing axonal sprouting, axonal route-finding, and synaptogenesis. As an illustration of underlying cortical histology during this period, the complex but ordered development of the human hippocampus proceeds from the primordial zonal arrangement (the ventricular neurogenesis source and intermediate zones, the neuronal migration target cortical plate, and outer neuron-poor marginal zone) (Noctor, Martinez-Cerdeno, & Kriegstein, 2007) at 9 weeks, through elaboration of subfields at 15-19 weeks, to morphologic maturity at 32-34 weeks gestation (Arnold & Trojanowski, 1996). The cytoarchitecture of the hippocampus is maintained after this time, although decreased neuronal density and enlargement of neurons occur throughout adolescence and beyond. Asynchrony of hippocampal complex development occurs, with the subiculum reaching maturity earliest, and the dentate latest in morphologic maturity. Myelenation only proceeds in this region in the weeks before birth, and continues through adolescence. This principal sequence of events and regional asynchrony is representative of other human cortical areas during brain development (Sidman & Rakic, 1982).

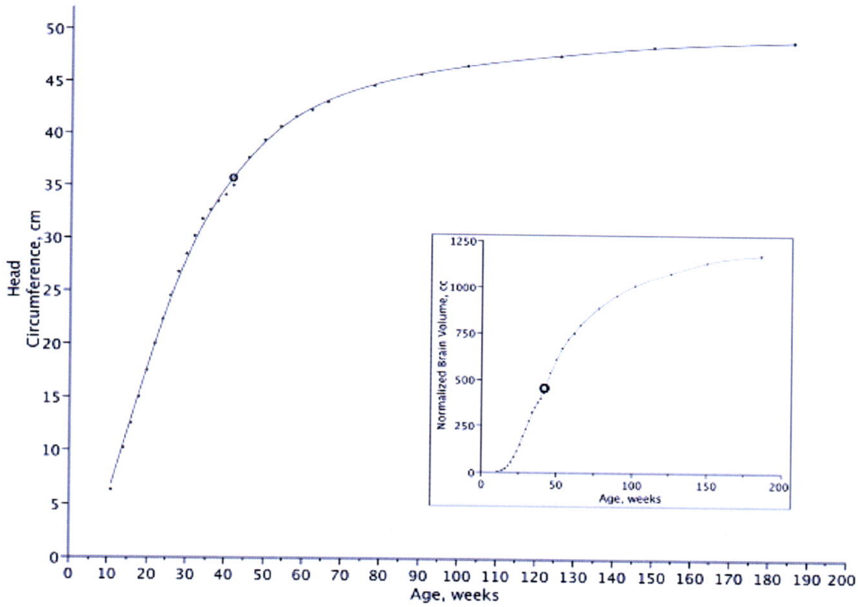

Fig. 1. Head circumference (HC) in centimeters from 10 weeks in utero (from Lessoway, Schulzer,Wittmann, Gagnon, & Wilson, (2008)) to 3 postnatal years, in weeks (from CDC growth charts, http://www.cdc.gov/nchs, 2007). There is approximately linear HC growth to at least 36 weeks gestation, corresponding in large part to neurogenesis and neuronal migration, and gradually diminishing postnatal HC growth associated with synaptogenesis and myelination processes. For continuity, the postnatal HC is averaged between boys and girls because the in utero results were not reported by gender. The curve is a spline smoothing fit to the 50th percentile data to show the trend; the apparent discontinuity at term could be due to a true slowing of head growth in the weeks just before birth, or differences in definition of gestational age, gender distribution, and technique of HC measurement between sources. Inset shows total volume (TIV) change estimated from HC, normalized by average TIV at term (427.4 cc;(Zacharia et al., 2006)). The TIV increases by approximately 11% between 3 and 16 years of age (Zhang et al., 2005). Circle (o) marks weeks at birth.

Only recently have brain imaging techniques allowed measurements of gray matter (GM) and white matter (WM) alterations with normal postnatal development. Gray matter density, used as a proxy for cortical developmental maturation, shows an increase before puberty followed by a decrease in adolescence and early adulthood that can be interpreted as a "pruning" process associated with reorganization and consolidation of efficient synaptic connections (Courchesne et al., 2000; Giedd et al., 1999; Jernigan & Tallal, 1990; Jernigan, Trauner, Hesselink, & Tallal (1991); Sowell, Thompson, Tessner, & Toga, 2001; Wilke, Krageloh-Mann, & Holland, 2007). This pruning process is invisible to head circumference measurements (Figure 1) because head circumference reflects maximum achieved brain size and would not decrease despite a small decrease with brain volume during adolescence. Cortical GM density maps based on longitudinal MRI data from normal subjects in the age range 4 – 21 years demonstrate early maturation (4 – 10 years) in the limbic and primary sensorimotor cortical regions, and later maturation (10 – 20 years) in modality-specific association and polymodal neocortex (Figure 2) (Gogtay et al., 2004). It is notable that even the primary regions showing the earliest and greatest reduction in GM density still retain the capacity for synaptic reorganization into adulthood (Ramachandran, 2005a, b).

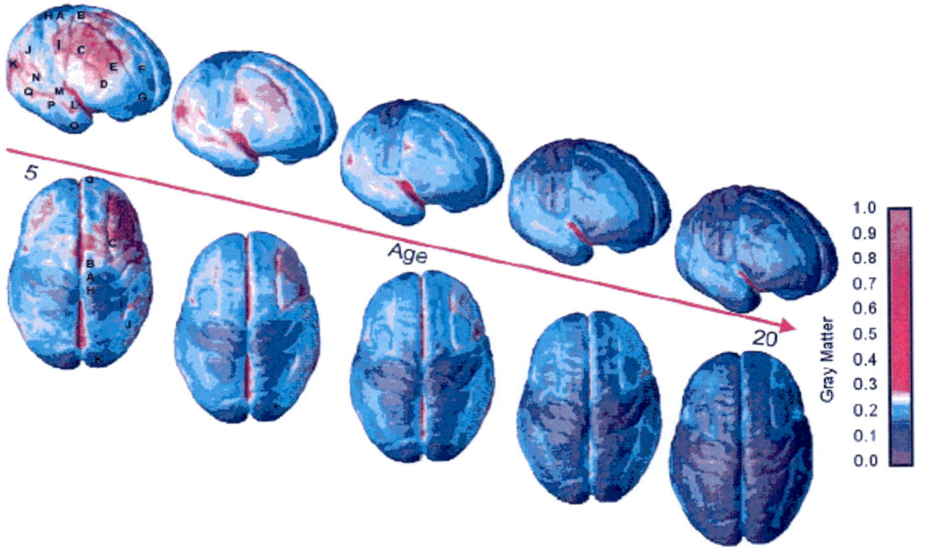

Fig. 2. Illustration from (Gogtay et al., 2004) showing progressive normal cortical maturation in 13 subjects scanned biennially from ages 4 to 21 years. Blue-green scale colors correspond to lower GM densities, red-yellow to higher densities. Early regional maturation is a pattern of decreasing GM density involving primary and polar neocortical and limbic (not shown) areas. Later GM maturation occurs in the premotor, modality-specific association, and polymodal neocortex.

Normal WM development in childhood and adolescence follows the sequence classically described by Yakovlev and Lecours (Yakovlev & Lecours, 1967), later observed using MRI (Barkovich, Kjos, Jackson & Norman, 1988). Myelination in the CNS proceeds from early maturation in basic motor and sensory systems in the brainstem, and begins initially in the corticospinal and primary sensory tracts and the optic tracts and radiations in the cerebral hemispheres. Myelination is synchronous with the cortical maturation pattern described earlier, with later stages of maturation occurring in modality-specific association and polymodal neocortex, particularly in the frontal lobe U-fibers. Association bundles, e.g., the corpus callosum, increase in relative size during the first 20 years of life, paralleling these later changes (Giedd et al., 1999; Rauch & Jinkins, 1994). Diffusion tensor MRI studies have confirmed and refined these observations (Schmithorst, Wilke, Dardzinski, & Holland, 2002), for example demonstrating regional gender-specific effects: females aged 5–18 years demonstrated greater age-related increases in fiber density (by mean diffusivity) than males in associative regions, whereas males had greater absolute organization of myelinated tracts (by fractional anisotropy) in the same regions (Schmithorst, Holland, & Dardzinski, 2007). The myelination process continues well into adulthood, for example in the hippocampal/parahippocampal region (Benes, Turtle, Khan, & Farol, 1994).

Functional studies to complement regional GM and WM developmental patterns have focused on language development. Word fluency tasks activate more widespread cortical areas, e.g., the right inferior frontal lobe in 11-year old children compared to 28 year-old adults. Language lateralization to a verb generation task likewise increases between 5 and 20 years of age (Szaflarski, Holland, Schmithorst, & Byars, 2006). In word-picture matching over a similar age range, shifts in activation were observed from frontal cortical regions to more posterior-inferior regions typical of the adult (Schmithorst, Holland, & Plante, 2006). A complex model for developmental changes in fMRI activation in a passive story-listening task has been presented, and generally supports a developmental shift in the strengths of functional connections between cortical language regions to the left hemisphere between 5 and 18 years of age (Karunanayaka et al., 2007).

The complex ordered developmental events just described provide a rich substrate for variable contributions to brain reserve. The numbers of neurons generated and their distribution through migration to form the cortical mantle, proper route-finding and establishment of synaptic connections between neurons, myelination of established fiber tracts, reinforcement of functional connections during the development of language and pruning of synapses in adolescence represent a few potential sources of these variations. These developmental events may therefore represent critical variables explaining different individual levels of resistance to the effects of brain pathology acquired years later in life.

2.2. Education and Cultural Influences

Education here refers to formal educational experiences, and cultural influences everything else, including family and friends, community, media, occupation, recreational activities and religion. Potential "biologic" exposures that might affect brain development or function, e.g., diet, water and air, toxin and drug exposures, and effects of medical disease are not considered.

The association between higher educational achievement and decreased risk of dementia is well established in the literature (Katzman, 1993). However, the level of education that best captures this association is controversial and an important issue because education before the end of adolescence is received in a critical developmental period biologically distinct from the adult, as described previously.

Some studies find the education effect only for no-education or primary school levels (Letenneur et al., 1999), but others find effects (odds ratio of about 2 for dementia) when defining low education as less than 8 years (Stern et al., 1994). Mixed model regression of education years in persons with pathologically defined AD but without clinical dementia showed that higher education was positively associated with this discordance (Roe, Xiong, Miller, & Morris, 2007). Expression of dementia in the presence of significant AD neuropathology was less likely by a factor of approximately 0.85 per year of additional education, supporting a brain reserve effect of education. An arbitrary education threshold was not used to produce this finding.

Related effects of education are that once symptoms begin, affected patients with higher educational attainment have greater atrophy (Kidron et al., 1997), greater metabolic abnormalities on functional imaging (Liao et al., 2005; Perneczky et al., 2006; Scarmeas et al., 2004; Y. Stern, Alexander, Prohovnik, & Mayeux, 1992), and greater cognitive decline to dementia (Unverzagt, Hui, Farlow, Hall, & Hendrie, 1998). These findings suggest that education may reduce or delay the expression of cognitive symptoms through a contribution to brain reserve, but once symptoms are expressed this reserve benefit is lost because the neuropathologic component of the underlying disease is more advanced at symptom onset.

Studies have shown that higher occupational attainment, which is difficult to disentangle from such factors as baseline intelligence, education and social advantage (Plassman et al., 1995), is associated with higher cognitive performance in late life and a lower likelihood of dementia (Andel et al., 2005; Potter, Plassman, Helms, Foster, & Edwards, 2006; Smyth et al., 2004), but again with an accelerated decline after symptoms begin (Andel, Vigen, Mack, Clark, & Gatz, 2006). Participation in cognitively stimulating activities may be the common link between brain reserve, education, and occupation (Wilson et al., 2002).

Recent studies have investigated the relationship of physical activity to cognitive performance, based in part on the hypothesis that such activity may stimulate adult neurogenesis and memory in mice (Steiner, Wolf, & Kempermann, 2006; Van der Borght, Havekes, Bos, Eggen, & Van der Zee, 2007). Although the contribution of physical fitness to overall health is unquestionable, the studies investigating a link of physical activity to brain reserve have been mixed (Kramer & Erickson, 2007; Scherder, Eggermont, Sergeant, & Boersma, 2007). Recent human and animal studies have not supported a contribution of "noncognitive" physical activity to cognitive impairment when environmentally stimulating cognitive activities are accounted for (Cracchiolo et al., 2007; Rovio et al., 2007). More work needs to be done in this area to supplement the education and occupation findings.

2.3. Genetic and Epigenetic Contributions to Brain Reserve

Here genetic refers to classic relationships between genotype and expression of traits and epigenetic denotes post-translational regulation and modification (e.g., DNA methylation) processes which may also alter that expression. The field of play for genetic and epigenetic mechanisms to influence brain reserve is quite large, including neurogenesis, neuronal migration and apportioning to functional regions, axonal route-finding and synaptogenesis, myelination, and synaptic modifications underlying particular types of neuronal plasticity. This section is restricted to an illustrative example of how these mechanisms could contribute to brain reserve. Neural plasticity is discussed further on in this review.

Migration of neurons from the ventricular zone to form the cortical mantle and the subsequent formation of the neocortical layers is a complex process whose genetic basis is just beginning to be understood (Rakic, 2007). This process occurs from 6-8 weeks gestation to four weeks before term (Sidman & Rakic, 1982). The migration process is guided by a specialized developmental scaffold consisting of the radial glia (Noctor, Martinez-Cerdeno, & Kriegstein, 2007; Rakic, 1990). The numerous anatomically distinct regions in the adult cortex are produced by preordained targeted neuronal migration to these regions and by continued elaboration of specific inter- and intra-regional connections (Rakic, 2007; Sidman & Rakic, 1982). With maturation, myelination patterns help to further define the adult architectonic areas (Bailey & Von Bonin, 1951; Brazier & Petsche, 1978). Genetic and epigenetic mechanisms orchestrate this process in part through determining neuronal numbers via mitotic division (Chenn & Walsh, 2002), motility patterns through calcium channel expression (Komuro & Rakic, 1996, 1998), filamin-A and MEKK4 (mitogen-activated protein kinase kinase kinase 4) signaling (Sarkisian et al., 2006), adhesion preference and strength via alpha3beta1 and alpha(v) integrins (Anton, Kreidberg, & Rakic, 1999), and termination of migration associated with expression of radial glial protein SPARC-like1 (secreted protein acidic and rich in cysteine-like 1) (Gongidi et al., 2004).

This process can go awry in association with human mutations in filamin-A, reelin (RELN), DCX, and aristaless-related homeobox protein (ARX), among many others (Barkovich, Kuzniecky, Jackson, Guerrini, & Dobyns, 2005). The complexity of interaction is illustrated by mutational variation in ARX, which has roles in several genetic cortical malformation disorders ranging from very mild to severe (Friocourt, Poirier, Rakic, Parnavelas, & Chelly, 2006; Kato et al., 2004; Stromme et al., 2002), and an important role in migration, neuronal proliferation, and maturation. Given the complexity of these multiple interactions in normal cortical development, a portion of brain reserve may well originate in genetic variations in ARX and other genes in the absence of mutation, influencing numerous critical points during cortical development, e.g., generation of raw primordial neuron numbers, development of and signaling by radial glia, neuronal place encoding in the ventricular and germinal matrix zones, targeting of neurons to their assigned region and layer of cortex, and formation of connections within and between cortical columns.

Neuropathology of AD acquired many years later develops within this field of genetic and epigenetically determined normal brain developmental variation. The amyloid precursor protein family, directly associated with AD neuropathology, also plays a key role in cortical development, particularly amyloid precursor like protein 1 (APLP1) (Lorent et al., 1995). Adaptor proteins interacting with this family, e.g. FE65, are required for normal cortical development (Guenette et al., 2006). Several other known or candidate genes in AD have important neurodevelopmental roles, particularly the presenilins and components of the Wnt signaling pathway, e.g., beta catenin (Chenn & Walsh, 2003; Ertekin-Taner, 2007; Malaterre, Ramsay, & Mantamadiotis, 2007; Wines-Samuelson & Shen, 2005). Recently the GRB-associated binding protein 2 (GAB2) gene has been found to be associated with late-onset AD, perhaps through interaction with other AD-related genes (E. M. Reiman et al., 2007). Much further work is needed to specify whether and to what degree genetic and epigenetic variation in specific pathways of brain development contribute to brain reserve and modification of symptom expression, apart from any role in AD neuropathology itself.

2.4. Normal Brain Alterations with Aging

Normal brain aging can be seen as consisting of consolidation, ongoing learning and adaptation on one hand, and erosion of brain reserve by involution on the other (Stern, 2002). On balance in successful brain aging, overall cognitive function is preserved. Here the focus is on structural anatomic alterations in the brain with normal aging.

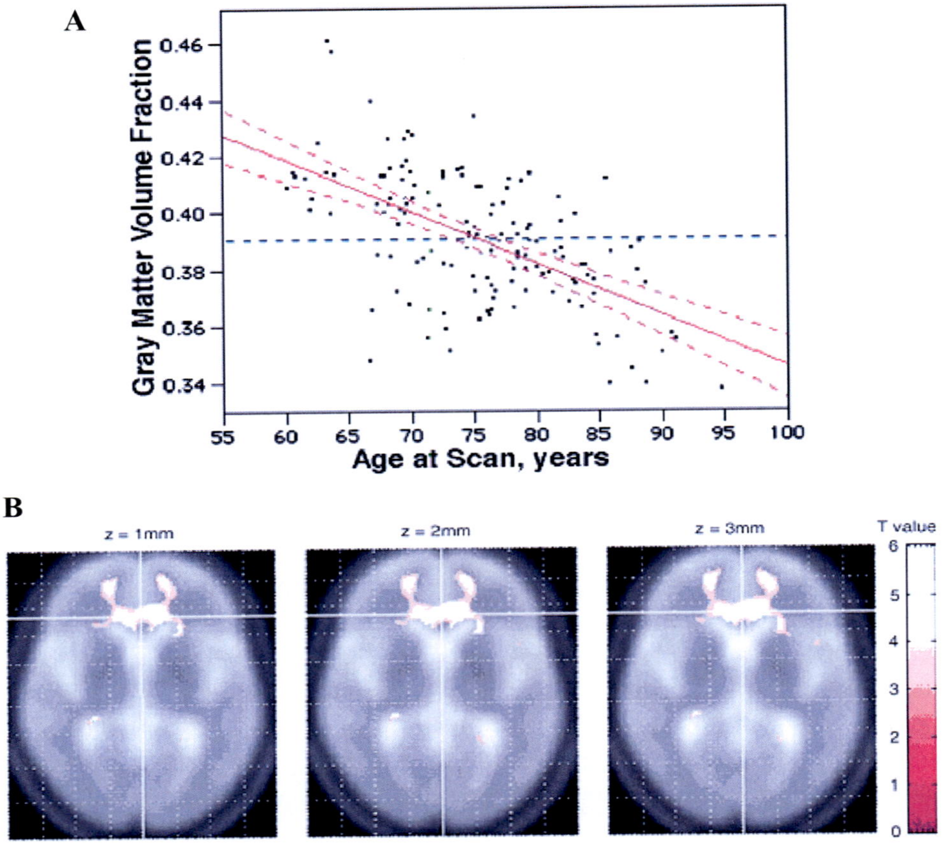

Fig. 3. (A). Decline in TIV-normalized gray matter volume (fraction) with age in 122 normal subjects aged 58 – 95 years. In this cross-sectional study, there was a corresponding increase in cerebrospinal fluid volume (CSF) with age, but no change in global WM volume (Smith, Chebrolu, Wekstein, Schmitt, & Markesbery, 2007). (B). Localized WM volume decreases were demonstrated in the corpus callosum genu and frontal lobe WM (significant decreases in yellow on ascending axial slices with Talairach coordinates shown (Smith, Chebrolu, Wekstein, Schmitt, & Markesbery, 2007)).

Although there are some studies that suggest a somewhat different pattern (Allen, Bruss, Brown, & Damasio, 2005; Liu et al., 2003), others have shown that after a period of relative stability from the early 20s, there is a slow reduction in brain volume with age beginning at approximately 50-55 years that accelerates after age 70 (Figure 3A) (Coffey, Saxton, Ratcliff, Bryan, & Lucke, 1999; Coffey et al., 1992; Courchesne et al., 2000; DeCarli et al., 2005; Jernigan & Gamst, 2005; Lemaitre et al., 2005; Lim, Zipursky, Watts, & Pfefferbaum, 1992;

Nagata et al., 1987; Pfefferbaum et al., 1994; Pfefferbaum, Sullivan, Swan, & Carmelli, 2000; Resnick, Pham, Kraut, Zonderman, & Davatzikos, 2003; Scahill et al., 2003). Voxel-based morphometric methods applied to magnetic resonance images have shown that GM volume and density reductions are global but the greatest reductions involve the parietal and frontal lobes (Good et al., 2001; Smith, Chebrolu, Wekstein, Schmitt, & Markesbery, 2007). Overall brain volume loss rates are on the order of 0.1 – 0.2% per year. There is little evidence of a selective regional volume loss exceeding global loss in the medial temporal lobe, the site of the earliest AD neuropathology. (DeCarli et al., 1994; Good et al., 2001; Smith, Chebrolu, Wekstein, Schmitt, & Markesbery, 2007). In this respect normal aging appears to differ from MCI and AD. It has been suggested that certain subcortical structures (substantia innominata) important in AD also undergo neuronal loss with aging, and that these losses may be partially reversed with intervention, e.g., nerve growth factor, at least in primates (Smith, Roberts, Gage, & Tuszynski, 1999).

The presence of global WM volume reductions with age remains an unresolved question, because some studies report it while others show no significant change (Good et al., 2001; Resnick et al., 2003; Smith, Chebrolu, Wekstein, Schmitt, & Markesbery, 2007; Walhovd et al., 2005). Other evidence suggest focal decreases in WM volume, specifically in the anterior corpus callosum and frontal lobe (Figure 3B) (Head, Snyder, Girton, Morris, & Buckner, 2005; Raz et al., 1997; Salat, Kaye, & Janowsky, 1999; Smith, Chebrolu, Wekstein, Schmitt, & Markesbery, 2007). However, the "frontal lobe hypothesis," used to explain declines in executive functions with age by positing a selective frontal involution, has been challenged (Greenwood, 2000).

The underlying morphologic correlates of brain volume reductions may include restricted neuronal losses, neuronal shrinkage, reduction in the size and extent of neuronal arbors and dendritic spines, and loss of myelin (Coleman & Flood, 1987; Dickstein et al., 2007; Flood & Coleman, 1988; Uylings & de Brabander, 2002). Therefore despite normal cognitive function, there may be diminished brain reserve measured generally by brain, and particularly cortical GM, volume loss in aging that may precede significant cognitive changes by many years. This cerebral atrophy is distinct from head circumference or total intracranial volume (TIV) as a measure of brain reserve (Schofield, Mosesson, Stern, & Mayeux, 1995; Wolf et al., 2004). Head circumference and TIV are better interpreted as a measure of in utero and postnatal brain developmental reserve (Bartholomeusz, Courchesne, & Karns, 2002).

2.5. Summary

Brain reserve is a useful general concept to help explain the ameliorating effects of education, occupation, and brain size on development of cognitive symptoms and dementia in late life. Specific potential components of this reserve discussed include the absolute number of neurons generated during early brain development, the targeted migration of these neurons to specific cortical regions, number and connections of cortical columns, editing and pruning of these connections during adolescence, and myelination of connecting fibers, most or all likely under genetic and epigenetic control. These components are consistent with known patterns of brain growth. Modifications of this substrate occur in the acquisition of language and under the influence of education, occupation, and other cognitive experiences that continue into adulthood.

Age-related decline in the morphologic characteristics and size or density of brain substance, particularly GM, is clearly established, and may diminish brain reserve in an age range when vulnerability to neurodegenerative pathologies is increased. The pattern of this age-related loss differs from the pattern seen in AD. Better understanding of the neurobiology of human brain development, even at its earliest stages, could provide important insights into how developmental processes leading to brain reserve could be enhanced. Currently little is known about specific contributions to the observed brain GM loss with normal aging, but there is no evidence that it is irreversible.

3. Neuronal Plasticity

The concept that the brain undergoes continuous modification and change, particularly at neuronal synapses, dates back inevitably to Cajal, who explicitly used a Spanish term equivalent to "neuronal plasticity" (Cajal, 1995). However, the dynamic and muscular nature of synaptic change was demonstrated much more recently (Buonomano & Merzenich, 1998). The discussion of this vast topic is restricted to modification of synaptic connections in adult cortex as a response to two circumstances: (1) learning, broadly defined and (2) compensation for injuries to the nervous system.

3.1. Synaptic Changes with Learning

Evidence of synaptic plasticity is based on two types of observations: (1) changes in paradigmatic stimulation-induced evoked responses such as long-term potentiation or depression, and (2) morphologic changes in axons, boutons and dendritic spines (Deng & Dunaevsky, 2005; Lippman & Dunaevsky, 2005; Stettler, Yamahachi, Li, Denk, & Gilbert, 2006; Tsanov & Manahan-Vaughan, 2007). Alterations in synaptic signaling and structural proteins underlie many of these changes (Ethell & Pasquale, 2005; Gomes, Hampton, El-Sabeawy, Sabo, & McAllister, 2006; Sharma, Fong, & Craig, 2006; Tao-Cheng, 2006). That local synaptic responses and morphology undergo change with learning, e g, exposure to different sensory inputs, in the adult brain is no longer in doubt (Buonomano & Merzenich, 1998). What remains at issue is how these changes are related to larger scale reorganization of cortical representations or maps, and whether alterations in these maps and their interconnections form the essence of learning and learning-induced changes in behavior and cognition. For the present the assumption will be that this is the case and that in the adult human cortex, there is an ongoing normal process of orderly turnover, re-tuning and remodeling of synapses that underlies brain adaptations to a changing internal and external environment, constituting learning.

The questions raised here are whether biologic variations in the quality and efficiency of neuronal plasticity constitute an important component of brain reserve, and whether age-related losses in that quality might erode that reserve. In the absence of an accepted measure of the quality of synaptic plastic processes that would encompass variations within the biologically normal range, specific answers are not currently available. However, there are several identified biochemical pathways that have been associated with neuronal plasticity and learning where such biologic variations are likely to occur (Figure 4).

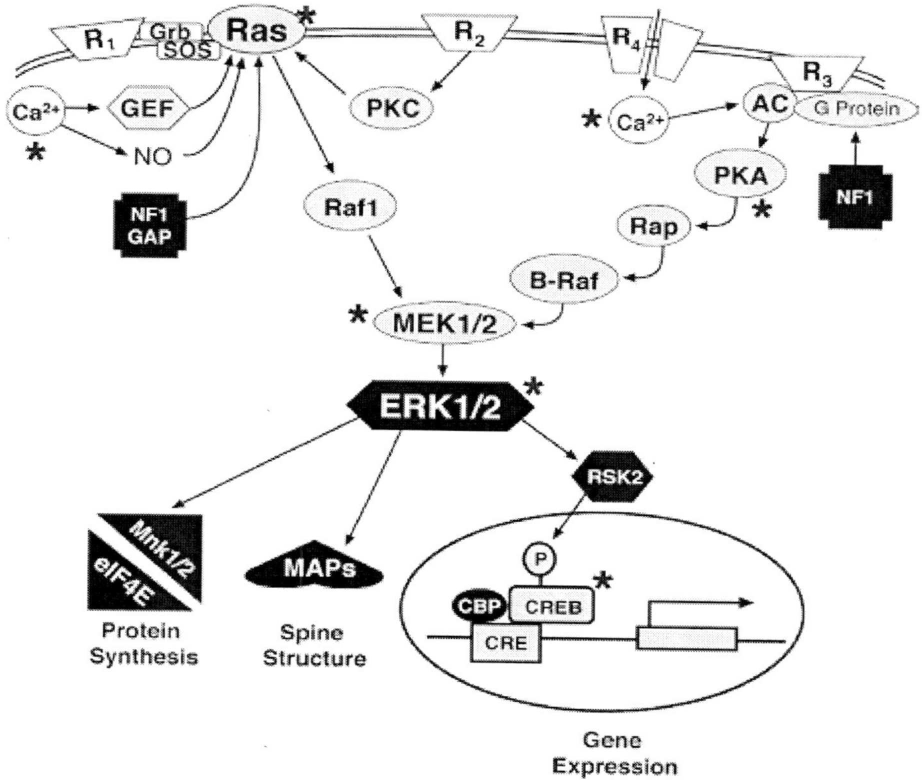

Fig. 4. Diagram of a current model of neuronal learning modified (from Sweatt et al., 2003). Key participants in the signaling cascades discussed in the text are indicated by an asterisk (*) in the figure: Ras, Calcium (Ca2+), PKA, MEK1/2, ERF1/2, and CREB. End points of the cascades are gene transcription and spine structural and other protein synthesis. The letter R denotes a receptor, some of which are associated with coupling proteins. Further details can be found in the reference.

A signaling pathway with a nuclear target and positive and negative regulatory elements essential for the formation of long term memory, involving protein kinase A (PKA) and the cyclic AMP-response element binding protein (CREB), is well established (Abel & Kandel, 1998; Silva & Giese, 1994). Continued research has demonstrated a mitogen-activated protein kinase (MAPK) pathway that involves extracellular signal regulated kinases (erk) 1 and 2, which couple to protein and ion channel expression mechanisms, but that also modulate CREB through phosphorylation. A complementary pathway headed by Ras is inhibited by PKA; the two paths overlap by modulating MAPK kinase (MEK1 and 2) (Lonze & Ginty, 2002; Sweatt, 2001; Sweatt, Weeber, & Lombroso, 2003; Waltereit & Weller, 2003). The dual role of MAPK in memory and growth is incorporated within the concept that synaptic plasticity involves both changes in synaptic potential responses and in synaptic shape and size – physiological and morphological plasticity (Thomas & Huganir, 2004). Recent studies show that CREB-mediated transcription may be governed by an "off-on switch" provided by the phosphorylation state of eukaryotic translation initiation factor 2, subunit 1 alpha (eIF2alpha) (Costa-Mattioli et al., 2007).

Other molecular biology experiments suggest that presenilins and presenilin-interacting proteins (delta-catenin) play important roles in normal synaptic plasticity, in part through interactions with Notch and CREB signaling pathways (Israely et al., 2004; Marjaux, Hartmann, & De Strooper, 2004; Saura et al., 2004; Wines-Samuelson & Shen, 2005). Amyloid precursor protein (APP) is likewise associated with normal plasticity and is cleaved by gamma-secretase, which requires presenilin for this activity (Gralle & Ferreira, 2007; Israely et al., 2004; Mileusnic, Lancashire, & Rose, 2005; Schmidt et al., 2007; Turner, O'Connor, Tate, & Abraham, 2003; Zhang et al., 2007). Mutations in the genes for presenilin and APP cause autosomal dominant early onset Alzheimer's disease, but this does not exclude a potential contribution to brain reserve in their normal roles in synaptic plasticity (Marjaux et al., 2004; Mileusnic et al., 2005). There is clearly ample range in these pathways for variations in synaptic plasticity that could explain normal adult differences in learning and in variable erosion of brain reserve with age.

Neuronal plasticity is a "driven" process depending on signals from sensory inputs and from other connected brain regions. The quality of those inputs in part derives from the integrity of the peripheral sensory organs and receptors and the complexity and nature of environmental stimuli encountered. Because both of these factors may change with age, it is difficult to distinguish a normal adaptation of a functioning plastic mechanism due to changes in these factors and age-related changes in the plastic mechanisms themselves. Both could produce functional and even macroscopic morphologic changes seen with age because neural plasticity has both physiologic and morphologic components as described. There is little experimental evidence bearing on this distinction (Disterhoft & Oh, 2006; Drapeau, Montaron, Aguerre, & Abrous, 2007; Mothet et al., 2006). Some established changes with aging could affect the intrinsic mechanisms of neuronal plasticity, e.g., calcium dysregulation, because calcium is the key activator of both the Ras and MAPK signaling cascades (Foster, 2007; Mattson, 2007). However, despite the likely loss of brain reserve capacity, there is evidence that learning remains largely intact in normal successful aging and may be enhanced by intervention (Mahncke et al., 2006).

3.2. Compensations in Brain Injury

This section briefly explores what potentials and limits there might be in the capacity of plastic changes in the adult to compensate for an underlying neurodegenerative pathology. Certain forms of compensation such as modification of alternative macroscopic networks are not considered.

The dynamic nature of cortical maps has been shown by overlaps and shifts in receptive fields for the remaining innervated skin following limb amputation in young adults, demonstrating a potential capacity in this form of adult macroscopic cortical plasticity (Ramachandran, 2005a; Ramachandran & Rogers-Ramachandran, 2000). There is now a large literature in "applied neuronal plasticity," the attempt to move basic ideas concerning plasticity to the clinic to benefit brain injured patients to rehabilitate and compensate these injuries (Butefisch, 2006). Various evidence-based approaches have focused on motor rehabilitation during intensive physical therapy, including increasing cortical excitability in the injured region using repetitive transcranial magnetic stimulation (Butler & Wolf, 2007; Mally & Stone, 2007), increasing input to damaged cortex through peripheral sensory nerve stimulation (Castel-Lacanal, Gerdelat-Mas, Marque, Loubinoux, & Simonetta-Moreau, 2007; Conforto, Cohen, dos Santos, Scaff, & Marie, 2007), immobilization (constraint) of the unaffected extremity (Mark, Taub, & Morris, 2006; Sunderland & Tuke, 2005), and enhancement of synaptic activation using noradrenergic and dopaminergic drugs, e.g., amphetamine (Adkins & Jones, 2005; Ramic et al., 2006; Walker-Batson et al., 2001). Some of these efforts have been successful, but effects of many of these therapies have not been systematically demonstrated in randomized trials. Proof that improvements observed are in fact due to alterations in neuronal plasticity is not yet available. Nonetheless the principle that neural plasticity can usefully compensate for brain injury is supported by early studies in this field. Perhaps further research in this area will provide greater insight into why head injury is associated with AD (Blasko et al., 2004; Ikonomovic et al., 2004; Olsson et al., 2004; Szczygielski et al., 2005).

A surprising result from such studies is that brain "compensation" could impede recovery in the injured brain region through cortical inhibition from remaining intact regions, or by the expression of proteins that inhibit neural outgrowth and other aspects of plasticity, e.g., the Nogo-A protein (Chen et al., 2000). These results suggest that brain compensations for injury may be quite different from simple enhancement of the routine expression of neuronal plasticity, and that some compensations may actually reduce it. Deeper understanding of these compensatory mechanisms may lead to therapies designed to "unlock" normal plastic mechanisms to promote recovery in injured areas, e.g., antibodies to Nogo (Emerick & Kartje, 2004; Emerick, Neafsey, Schwab, & Kartje, 2003; Markus et al., 2005; Papadopoulos et al., 2002). There is a potential for such therapies in neurodegeneration if similar compensatory changes occur in those conditions, an unexplored possibility at present. Other compensatory mechanisms have been suggested to explain variability in Down's syndrome dementia, a dementia associated with AD neuropathology (Head, Lott, Patterson, Doran, & Haier, 2007).

3.3. Summary

Neural plasticity is the dominant known substrate for learning in the adult and has a detailed physiologic, morphologic, and molecular underpinning. Presenilin and APP have important normal roles in this substrate. Normal biologic variations in the quality and efficiency of plastic processes can plausibly explain differences in learning capacity between normal adults, a contribution to variations in brain reserve. Age-related decline in plasticity could be due to normal neural plastic responses to changes in sensory and environmental stimuli with age, age-related alterations in the character of plastic mechanisms themselves, or both. This decline appears small in relation to preservation of learning in aging, but it may signal the presence of an underlying erosion in brain reserve that can later be exposed by developing neuropathologic alterations in synapses. The brain compensates for injury at least in part through mechanisms of neural plasticity, but in some circumstances plasticity in the injured region may actually be reduced.

4. Mild Cognitive Impairment (MCI) and Alzheimer's Disease (AD)

A simple listing of topic areas in this huge literature and its controversies would take many pages; therefore, simplification of complex issues is inevitable. Here the focus will be on key concepts related to MCI and AD that provide a basic foundation for the following section on presymptomatic detection.

4.1. Clinical Approach and Definition of MCI

Flicker et al showed progression of dementia in a group of longitudinally followed elderly subjects who initially had mild impairment on psychometric tests compared to controls (Flicker, Ferris, & Reisberg, 1991). Petersen et al emphasized and refined this concept of MCI and distinguished the amnestic form from other types (Petersen et al., 1999). The criteria of Petersen et al for amnestic MCI are most frequently used (Petersen, 1998): (a) memory complaints, preferably corroborated by an informant; (b) psychometrically defined memory impairment for age and education, but with preserved general cognitive function; (c) intact activities of daily living (ADLs); and (d) no clinical dementia. Subjects with amnestic MCI are by definition not demented, but are at greatly increased risk for the future development of dementia and in particular AD and vascular dementia, although rates of progression vary considerably across studies (Gauthier et al., 2006; Yaffe, Petersen, Lindquist, Kramer, & Miller, 2006).

The clinical approach is embodied in the clinical dementia rating (CDR), a formal structured interview of a patient and an informant by an experienced medical professional in which the opinion of degree of impairment is rated in five domains together with an overall rating score ranging from 0 to 3. In the proper context, this rating is highly reliable in the diagnosis and staging of MCI and dementia (Burke et al., 1988; Fillenbaum, Peterson, & Morris, 1996; Morris, 1997; Morris et al., 1991; Storandt, Grant, Miller, & Morris, 2006). A person who is asymptomatic by definition has a CDR of zero.

4.2. Neuropathology of MCI and AD

The diagnostic hallmarks of AD pathology are neuritic amyloid plaques, neuropil threads, and neurofibrillary tangles (Price et al., 1998). Neuronal and particularly synaptic losses accompany these hallmarks (Brun, Liu, & Erikson, 1995; P. Coleman, Federoff, & Kurlan, 2004; Masliah, Terry, DeTeresa, & Hansen, 1989; Scheff & Price, 2006; Scheff, Price, Schmitt, & Mufson, 2006; Small, 2004; Terry, 2006). The distribution of tangle neuropathology is incorporated into the Braak staging system for AD, which recognizes the earliest pathologic alterations in the transentorhinal cortex, followed by progressively greater tangle pathology in limbic regions, culminating in late stage widespread neocortical involvement (Braak, Alafuzoff, Arzberger, Kretzschmar, & Del Tredici, 2006; Braak, Braak, & Bohl, 1993). The rate of accumulation of this AD pathology is difficult to measure because in autopsy or biopsy only one time point can be sampled. Some studies have estimated a slightly increasing accumulation rate with age but the form of this acceleration is not clear, e.g., a power law or exponential rate increase (Dani, Pittella, Boehme, Hori, & Schneider, 1997; Miyasaka et al., 2005; Ohm, Muller, Braak, & Bohl, 1995).

In the presence of a clinical diagnosis of dementia and any acknowledged confounding brain pathologies, a neuropathologic diagnosis of AD can be made as low, intermediate, or high likelihood according to the National Institute on Aging and Reagan Institute criteria, which use semiquantitative estimates of CERAD neuritic plaque score and Braak staging. The stringency of these criteria depends on whether they are used for validation or other research, or for routine diagnostic purposes ("Consensus recommendations for the postmortem diagnosis of Alzheimer's disease. The National Institute on Aging, and Reagan Institute Working Group on Diagnostic Criteria for the Neuropathological Assessment of Alzheimer's Disease," 1997; Geddes et al., 1997; Jellinger, 1998; Jellinger & Bancher, 1998; Newell, Hyman, Growdon, & Hedley-Whyte, 1999; Wisniewski & Silverman, 1997). Neuropathology in amnestic MCI diagnosed in expert centers strongly resembles AD pathology in every respect except for lower levels of intensity and extent, supporting the construct of amnestic MCI as a symptomatic stage prior to clinically diagnosed AD (Bennett et al., 2006; Markesbery et al., 2006; Morris & Price, 2001; Petersen et al., 2006; Schmitt et al., 2000).

4.3. Non-AD Neuropathologies

The recognition of the contribution of non-AD neuropathologies to dementia, e.g., micro- and small infarcts and Lewy bodies, has complicated neuropathologic diagnosis (Kovari et al., 2007). Coexistence of these pathologies with AD pathology increases the likelihood of association with cognitive symptoms and clinical dementia diagnosis (Bennett, Schneider, Bienias, Evans, & Wilson, 2005; Schneider, Arvanitakis, Bang, & Bennett, 2007; Snowdon et al., 1997). Because Lewy body pathology and AD pathology frequently coexist at autopsy, diagnosis of a Lewy body component distinct form AD remains a difficult clinical challenge (Metzler-Baddeley, 2007; Weisman et al., 2007).

An important theoretical issue is whether brain reserve should be considered in relation to one particular pathology or to all pathologies combined. If in relation to one pathology such as AD, remaining pathologies could be seen as eroding brain reserve relative to AD. However, simply because some pathologies, e.g., small infarcts, are common in aging does not mean they should be swept into the concept of reserve. This point is emphasized if one considers also including asymptomatic Lewy body pathology as eroding brain reserve relative to AD, which appears unwise because Lewy body pathology alone can cause dementia (Marui, Iseki, Kato, Akatsu, & Kosaka, 2004). Perhaps it is better is to consider brain reserve in relation to accumulating neuropathology whether individual or combined and assessing contributions from each type.

4.4. The Amyloid Hypothesis and Disease-Modifying Treatment

The amyloid hypothesis takes several forms, the gist of which is that abnormally increased production of particular beta-amyloid cleavage fragments of APP leads to a cascade of damaging effects on brain synapses and neurons (Walsh & Selkoe, 2004). The eventual consequence of this damage is MCI and ultimately, AD. The scientific support for this hypothesis would fill many volumes.

Several potential pharmacologic treatments for AD based on the amyloid hypothesis are in development, two of which have completed or will soon complete phase-III clinical trials seeking a Food and Drug Administration indication for disease modification. Tramiprosate binds beta amyloid and is thought to prevent formation of amyloid fibrils, and tarenflurbil is a selective amyloid beta 1-42 lowering agent (Gervais et al., 2007; Golde, 2006; Kukar et al., 2007). None of the medications approved beginning with tacrine in 1993 have had such an indication, so the introduction of these therapies is eagerly awaited by AD patients and their physicians. Neither drug has known effects on brain reserve or on non-AD pathologies.

4.5. Summary

Research has resulted in reliable clinical criteria for the diagnosis of amnestic MCI, which conceptually represents the earliest symptoms of mild memory loss presaging the onset of AD. Individuals with amnestic MCI have established neuropathology typical of AD, but the intensity and extent of that pathology is generally less than in AD, suggesting that further progression of pathology accompanies the further progression of mild memory symptoms to dementia. Disease modifying treatment is in clinical trial and is expected to slow the progression of pathology and therefore the rate of clinical progression from MCI to AD. The presence of other pathologies increases the likelihood that a given level of AD pathology will be associated with dementia. Disease modifying treatment would not decrease this association but could reduce the level of AD pathology and consequent dementia.

5. Presymptomatic Detection of MCI and AD

The purpose of this section is to discuss evidence that AD pathology is present in normal persons for many years prior to the period of vulnerability for late-onset AD. Potential imaging, biofluid, clinical and psychometric methods for detecting effects of this pathology and predicting onset of early AD symptoms are presented.

5.1. Evidence for Presymptomatic Neuropathology in AD

Neurofibrillary pathology characteristic of AD but restricted to entorhinal and hippocampal areas (Braak stages I & II) was present in over 50% of autopsies performed between ages 50 and 60 years as part of a larger study of 2,661 staged neuropathologic examinations in the age range 25 – 95 years (Braak & Braak, 1998). The age range of 50-60 is below the usual expected onset of clinical late-onset AD (McKhann et al., 1984). The presence of AD pathology in older, carefully evaluated nondemented subjects has now been documented in several subsequent studies (Bennett et al., 2006; Davis, Schmitt, Wekstein, & Markesbery, 1999; Driscoll et al., 2006; Markesbery et al., 2006; Schmitt et al., 2000). Two key issues raised by these findings are: (1) whether this pathology is related to subtle cognitive decline to levels that remain within the normal range and that are insufficient to produce functional impairments, and (2) the characteristics of AD pathologic progression demarking the transition to the symptomatic state, e.g., MCI. The issue of cognitive change is discussed later in this section; here the neuropathologic correlates of transition to MCI are briefly discussed. .

Recent studies have shown that likelihood of early memory and related symptoms is related to two basic factors: the presence of coexistent non-AD pathologies, e.g., microinfarcts, and the extent or stage of AD pathology (Bennett et al., 2006; Bennett et al., 2005; Petersen et al., 2006; Schmitt et al., 2000; Schneider et al., 2007). Increased numbers of limbic neurofibrillary tangles and the presence of neocortical neuritic plaque pathology appear to best distinguish persons meeting clinical criteria for MCI from normal (Markesbery et al., 2006; Morris et al., 1991; Morris & Price, 2001). Nonetheless similar mild to moderate levels of AD pathology seen in MCI are also observed in well characterized subjects with normal levels of cognition and functional performance. The concept of brain reserve and its several components as variables explaining this discrepancy have been discussed in a previous section.

5.2. Imaging Biomarkers

Development of imaging as a technique to assess the brain in normal subjects prior to MCI and dementia may have different goals that should be distinguished: (1) detecting the presence of morphologic and functional changes due to the presence of underlying asymptomatic AD and other pathologies, (2) characterizing brain features related to brain reserve and changes in that reserve, and (3) mapping brain compensatory changes related to advancing neuropathology or to involutional age-related erosion of brain reserve. The potential for confounding these different goals depending on limitations of imaging techniques or of practical experimental designs should be kept in mind.

Much imaging research has been based on the hypothesis that because the earliest pathologic alterations in AD involve the entorhinal cortex and other medial temporal limbic structures, these regions will be the most salient in predicting normal subjects who would later develop MCI or AD (Kaye et al., 1997). Absolute entorhinal and hippocampal volumes are decreased and rates of volume change are increased in MCI (Convit et al., 2000; Jack et al., 1999; Jack et al., 2004; Xu et al., 2000). However, some studies have shown that general measures of brain atrophy rate are as salient as measures of entorhinal and hippocampal atrophy rate in predicting conversion from normal to MCI (Jack et al., 2004; Jack et al., 2005) or cognitive decline without dementia (Adak et al., 2004), suggesting that there may be a separate brain reserve component outside the medial temporal target region of early AD neuropathology (Chetelat et al., 2005; Convit et al., 2000; Jack et al., 2000; Karas et al., 2004; Mueller et al., 1998; Pennanen et al., 2005). Others have demonstrated decreased GM density in cognitively normal apolipoprotein (APOE) E4 allele carriers at risk for AD (Wishart, Saykin, McAllister et al., 2006).

A recent study demonstrated that medial temporal and left parietal GM volumes in normal subjects predicted MCI within five years with 76% accuracy that was further enhanced to 87% by combining GM volume with a cognitive measure, raw Wechsler Memory Scale score (Figure 5A) (Smith, Chebrolu, Wekstein, Schmitt, Jicha et al., 2007).

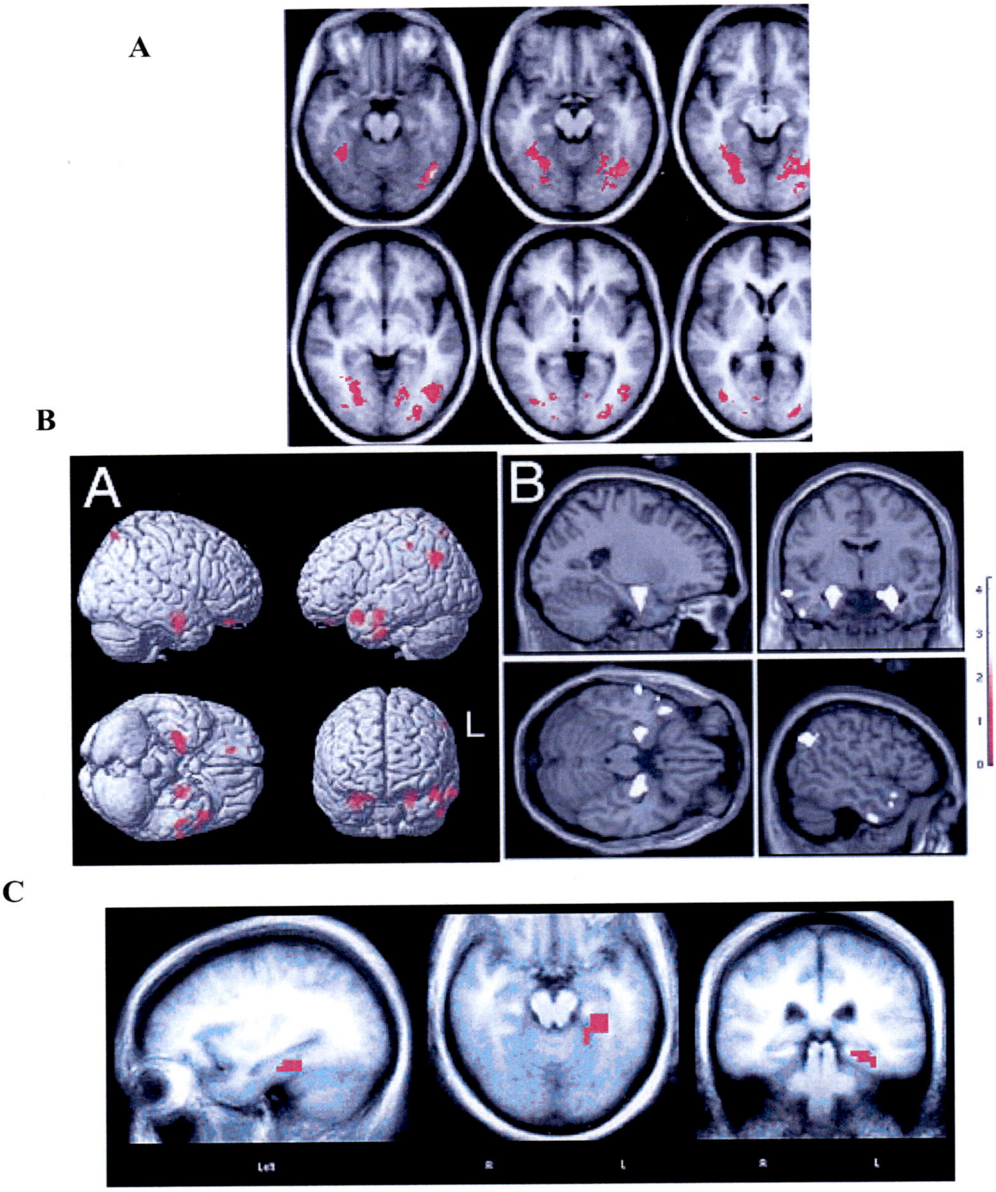

Fig. 5. Illustration of structural and functional imaging findings in asymptomatic normal subjects at risk of AD. (A) Regions of significant volume decrease in the anteromedial temporal lobes and left angular gyrus region in 23 normal subjects destined to develop MCI vs. 113 normal subjects who remained normal over 5 years follow-up assessment (Smith, Chebrolu, Wekstein, Schmitt, Jicha et al., 2007). **(B)** Decreased ventral temporal activation during picture naming in 14 presymptomatic normal subjects at increased risk of AD vs. 12 low-risk subjects (Smith et al., 1999). **(C)** Left medial temporal activation increase during paired word associates learning in 10 presymptomatic normal subjects at increased risk of AD vs. 10 low-risk subjects (Fleisher et al., 2005). In (A) & (B) the red and yellow overlay indicates a decrease in the comparison; in (C) red indicates an increase.

Lower resting glucose metabolism using positron emission tomography (PET) in the left angular, mid temporal, and mid frontal gyri predict decline on a global cognitive measure, the

Mini Mental State Examination, and decreased medial temporal volumes predict decline on delayed memory in normal subjects at baseline followed subsequently over 3.8 years (Jagust et al., 2006). Ten percent of these subjects developed MCI or dementia, but the correlation remained significant without them, suggesting these imaging measures may predict MCI or dementia in normal subjects.

Functional imaging studies have shown alterations in brain activation in subjects at increased risk of AD years before the onset of cognitive symptoms. Decreased PET resting glucose metabolism in the posterior cingulate and parietal regions has been demonstrated in normal subjects at increased risk of early-onset AD due presenilin and APP mutations, and in subjects with a family history of late-onset AD and at least one apolipoprotein (APOE) E4 allele (Johnson et al., 2001; Reiman et al., 1996; G. W. Small et al., 2000; Stefanova et al., 2002). Amyloid imaging, a PET-based technique for quantitating the burden of brain amyloid, is conceptually appealing and holds promise for presymptomatic detection (Lockhart et al., 2007; Nordberg, 2007; Rowe et al., 2007). The role of this new technique for presymptomatic detection is not yet established.Active functional MRI and PET results have not been entirely consistent. Some studies have demonstrated decreased activation in prefrontal activity in normal high-risk (APOE4 carrier) subjects during working memory tasks, in ventral temporal regions during picture naming, and in medial temporal regions during episodic memory encoding (Figure 5B) (Elgh, Larsson, Eriksson, & Nyberg, 2003; C. D. Smith et al., 1999; C. D. Smith et al., 2005; Trivedi et al., 2006). Other studies have demonstrated increased parietal activation associated with verbal fluency tasks and increased medial temporal activation during episodic memory encoding or retrieval (Figure 5C) (Bondi, Houston, Eyler, & Brown, 2005; Bookheimer et al., 2000; Fleisher et al., 2005; Smith et al., 2002; Sperling, 2007; Wishart, Saykin, Rabin et al., 2006). These functional studies differ in the age of subjects, inclusion of family history with the APOE4 carrier state, nature of tasks performed during imaging, thoroughness of cognitive testing, and methods of analyses. Nonetheless, these studies suggest that active functional imaging holds promise as class of imaging biomarkers for presymptomatic brain alterations in normal persons at risk for AD.

5.3. Serum and CSF Biomarkers

The ratio between beta amyloid isoforms 1-42 and 1-40 in plasma and CSF has been shown to associate with or predict risk of later MCI or AD in normal subjects in separate studies (Fagan et al., 2007; Graff-Radford et al., 2007). However, the relationship between plasma and CSF amyloid has not yet been fully clarified (Freeman, Raju, Hyman, Frosch, & Irizarry, 2007). A combination of tau, phospho-tau, and beta amyloid 1-42 CSF levels may improve prediction of conversion from MCI to AD. Other approaches to improve sensitivity and specificity of CSF analytes have included measurement of specific oxidation products, e.g. 4-hydroxynonenal, and protein patterning to detect a profile of neuropathology-related molecular alterations (Finehout, Franck, Choe, Relkin, & Lee, 2007; Lovell, Xie, & Markesbery, 1998). Studies have not yet examined these or similar promising CSF biomarker approaches to predict later cognitive symptoms of MCI in normal subjects.

5.4. Psychometric Assessments to Predict MCI or AD

That cognitive performance in early adulthood may predict performance in late life would not be surprising. For example, performance on a military examination, the Army General Classification Test accounted for 21% of the variance in cognitive performance measured 50 years later and education accounted for 16.7% (Plassman et al., 1995). However, cognitive performance in childhood and early adulthood has also been linked to later expression of dementia symptoms. Performance of 11-year-olds on a standardized mental ability test in Scotland in 1932 was lower in survivors who developed dementia after age 65 years than in those who did not. In a classic longitudinal study of nuns, idea density and grammatical complexity judged from essays written when subjects entered their religious order at age 22 was lower in those who developed dementia many years later (Snowdon et al., 1996). Investigators in the Framingham study found that derived memory scores were lower in normal subjects who developed dementia an average of 22 years later (Elias et al., 2000). These early detected differences in performance and ability are most likely determined by developmental differences discussed previously under the heading of brain reserve.

The literature on presymptomatic prediction is confounded by newly improved concepts and clinical techniques for the detection of MCI, which themselves include psychometric assessment as one of the criteria for this diagnosis. This moving target makes it more difficult to interpret literature demonstrating presymptomatic cognitive declines, particularly on subtle memory measures (Blacker et al., 2007; Crystal et al., 1996; Grober, Lipton, Hall, & Crystal, 2000; Masur, Sliwinski, Lipton, Blau, & Crystal, 1994; Small, Stern, Tang, & Mayeux, 1999; Tierney, Yao, Kiss, & McDowell, 2005). There is no doubt that decline on psychometric measures occurs prior to dementia.

The clinical and neuropsychological approaches to presymptomatic detection encompass a range: the CDR is a structured clinical evaluation that does not include formal psychometric testing (Morris, 1997); the diagnosis of MCI may include a CDR-like clinical assessment but also explicitly incorporates psychometric testing results within its criteria (Kelley & Petersen, 2007), and finally, many studies used cognitive assessments alone to determine pre-dementia states (Blacker et al., 2007). The types of evaluations bracketing the two ends of this range are fundamentally different and each approach has advantages and defects. The CDR is an individual determination that does not require adjustments for age or education (norms), but it does require clinical expertise, judgment, and time (Morris, 1997). The recently developed AD8 may overcome some of this limitation (Galvin, Roe, Xiong, & Morris, 2006). Psychometric evaluations provide quantitative results in detailed cognitive domains and are easier to implement in large-scale studies but scores must be adjusted for several confounds, including age, and may be subject to greater individual variation over time (Grober et al., 2000; Small et al., 1999; Tierney et al., 2005).

MCI diagnosis is the middle way, incorporating advantages of both approaches. However, in this case there is more data to consider, which may lead to the problem of reconciling the two when results conflict. One group that has used both methods independently reports no psychometrically-defined cognitive impairment or decline in longitudinally followed subjects prior to clinical detection, even in the presence of AD pathology (Goldman et al., 2001; Storandt et al., 2006). Others have drawn similar conclusions (Driscoll et al., 2006). This suggests that other methods are needed to detect AD pathology before MCI (Morris & Price, 2001).

5.5. Summary

AD neuropathology and often other pathologies are already present when the earliest memory and related cognitive symptoms first appear. In many cases these symptoms herald a progressive clinical dementia. These observations imply that neuropathology was present for some interval when the affected individual was cognitively and behaviorally indistinguishable from other normal persons without AD pathology. Accumulating data suggest this interval can be measured in decades. Current evidence indicates that the complementary clinical and psychometric detection methods available today may not make this critical distinction. If disease modifying treatments soon to be available slow the progress of AD pathology, it follows that these treatments could delay the onset of symptoms or perhaps prevent their appearance entirely. Imaging biomarkers and serum and CSF analytes may provide the means to detect presymptomatic pathology and follow its evolution to measure treatment effects in future preventive clinical trials.

6. Overall Summary and Conclusions

6.1. The Argument for Presymptomatic Detection

Here the use of qualifiers, e.g., "might", and "may", is dropped to emphasize the fundamental argument. Presymptomatic detection refers specifically to detection of late-onset AD.

Intrinsic or acquired brain properties confer variable resistance to the effects of neuropathology, modifying whether and when that neuropathology is expressed as symptoms, changes in cognitive performance, or behavior. These properties include developmental factors such as original neuronal number and the extent of synaptic and fiber pruning during adolescence. Rates of formation and maintenance of synaptic connections and of myelination of nerve fibers are critical sources of normal variation both in development and the adult. Educational and other sources of cognitive stimulation also contribute to brain reserve by increasing the redundancy or multiplicity of brain associative connections formed during childhood and throughout adulthood. Genetic and epigenetic mechanisms underlie these developmental events and the synaptic and neuronal molecular mechanisms of learning in the adult, constituting a source of normal biologic variation in reserve. Genes for APP and presenilin known to cause early-onset AD when mutated are essential to this variation. As the brain ages, there is a slow loss of global GM volume and a focal loss of frontal WM, eroding this reserve. Normal intrinsic neuronal plasticity compensates for a portion of this erosion and, in successful aging, cognition is preserved into late life.

AD neuropathology is present in significant amounts by at least age 50 years, and increases in intensity and extent despite preservation of cognition at this age. Imaging evidence demonstrates presymptomatic alterations in brain function in this age range in persons at increased risk of AD, representing brain compensation for the presence of AD pathology. Typically after age 65, whether and when cognitive symptoms appear in persons who have a given level of pathology is determined by: (1) the variable level of reserve set by its several underlying mechanisms or (2) exhaustion of existing compensation, or both. The

AD neuropathologic process itself may accelerate at this threshold. In this view, the appearance of symptoms marks a critical point where the future clinical trajectory is determined by AD and other pathologies; intrinsic reserve and compensation can no longer have significant effects. Even in the ideal circumstance where AD pathology is halted completely by specific disease-modifying treatment, other accumulating pathologies and age-related loss of brain substance can cause worsening.

The situation with disease-modification in the presymptomatic period is different. Slowing of the AD pathologic process before symptoms appear delays their appearance even if mild, allows compensatory mechanisms to address a slower progression of pathology, and saves the reserve needed to offset the effects of other pathologies, e.g., microinfarcts. If neuropathology accelerates with age or onset of symptoms, early intervention has compound effects.

6.2. Future Directions

The program suggested to address these conclusions is to focus more attention on presymptomatic detection based on the promising early results in this field. We need to know more about human developmental events and their molecular underpinning from the perspective of their potential effect in later life. In particular, the role of genes associated with early and late-onset AD, e.g., APP, presenilin, SORL1 and GAB2 in development should be further explored. We need to know more about the molecular events involved in normal adult synaptic plasticity and in related compensatory mechanisms, e.g., No-goA and the AD-associated genes. Further longitudinal imaging and clinical study of normal aging with continued follow-up into the MCI transition is needed to discover alterations prior to symptoms that are invisible to present clinical methods.

Excellent and laudable efforts are ongoing in many of these areas, but an organized focus from the perspective of late-life dementia is not always evident. Understandably much previous work has been devoted to characterizing symptomatic AD states, but some of the many brilliant minds behind these investigations could also advance the cause of presymptomatic detection for the reasons given here.

Acknowledgements

Apologies in advance for any omissions of research contributions not cited here; there is no intent to neglect or inaccurately represent anyone's work. Gratitude is owed to the many brilliant investigators who have made important contributions to AD research so that patients and families may benefit. The author thanks Dr. W.R. Markesbery for his mentorship, and the NIH for support through NIH grants P30 AG028383 and R01 NS36660.

References

Abel, T., & Kandel, E. (1998). Positive and negative regulatory mechanisms that mediate long-term memory storage. *Brain Res Brain Res Rev*, 26(2-3), 360-378.

Adak, S., Illouz, K., Gorman, W., Tandon, R., Zimmerman, E. A., Guariglia, R., Moore, M. M., & Kaye, J. A. (2004). Predicting the rate of cognitive decline in aging and early Alzheimer disease. *Neurology*, 63(1), 108-114.

Adkins, D. L., & Jones, T. A. (2005). D-amphetamine enhances skilled reaching after ischemic cortical lesions in rats. *Neurosci Lett*, 380(3), 214-218.

Allen, J. S., Bruss, J., Brown, C. K., & Damasio, H. (2005). Normal neuroanatomical variation due to age: the major lobes and a parcellation of the temporal region. *Neurobiol Aging*, 26(9), 1245-1260; discussion 1279-1282.

Andel, R., Crowe, M., Pedersen, N. L., Mortimer, J., Crimmins, E., Johansson, B., & Gatz, M. (2005). Complexity of work and risk of Alzheimer's disease: a population-based study of Swedish twins. *J Gerontol B Psychol Sci Soc Sci*, 60(5), P251-258.

Andel, R., Vigen, C., Mack, W. J., Clark, L. J., & Gatz, M. (2006). The effect of education and occupational complexity on rate of cognitive decline in Alzheimer's patients. *J Int Neuropsychol Soc*, 12(1), 147-152.

Anton, E. S., Kreidberg, J. A., & Rakic, P. (1999). Distinct functions of alpha3 and alpha(v) integrin receptors in neuronal migration and laminar organization of the cerebral cortex. *Neuron*, 22(2), 277-289.

Arnold, S. E., & Trojanowski, J. Q. (1996). Human fetal hippocampal development: I. Cytoarchitecture, myeloarchitecture, and neuronal morphologic features. *J Comp Neurol*, 367(2), 274-292.

Bailey, P., & Von Bonin, G. (1951). *The isocortex of man*. Urban: University of Illinois Press.

Barkovich A.J., Kjos, B. O., Jackson, D. E., & Norman, D. (1988). Normal maturation of the neonatal and infant brain: MR imaging at 1.5 T. *Radiology* 166, 173-180.

Barkovich, A. J., Kuzniecky, R. I., Jackson, G. D., Guerrini, R., & Dobyns, W. B. (2005). A developmental and genetic classification for malformations of cortical development. *Neurology*, 65(12), 1873-1887.

Bartholomeusz, H. H., Courchesne, E., & Karns, C. M. (2002). Relationship between head circumference and brain volume in healthy normal toddlers, children, and adults. *Neuropediatrics*, 33(5), 239-241.

Benes, F. M., Turtle, M., Khan, Y., & Farol, P. (1994). Myelination of a key relay zone in the hippocampal formation occurs in the human brain during childhood, adolescence, and adulthood. *Arch Gen Psychiatry*, 51(6), 477-484.

Bennett, D. A., Schneider, J. A., Arvanitakis, Z., Kelly, J. F., Aggarwal, N. T., Shah, R. C., & Wilson, R. S. (2006). Neuropathology of older persons without cognitive impairment from two community-based studies. *Neurology*, 66(12), 1837-1844.

Bennett, D. A., Schneider, J. A., Bienias, J. L., Evans, D. A., & Wilson, R. S. (2005). Mild cognitive impairment is related to Alzheimer disease pathology and cerebral infarctions. *Neurology*, 64(5), 834-841.

Blacker, D., Lee, H., Muzikansky, A., Martin, E. C., Tanzi, R., McArdle, J. J., Moss, M., & Albert, M. (2007). Neuropsychological measures in normal individuals that predict subsequent cognitive decline. *Arch Neurol*, 64(6), 862-871.

Blasko, I., Beer, R., Bigl, M., Apelt, J., Franz, G., Rudzki, D., Ransmayr, G., Kampfl, A., & Schliebs, R. (2004). Experimental traumatic brain injury in rats stimulates the expression, production and activity of Alzheimer's disease beta-secretase (BACE-1). *J Neural Transm*, 111(4), 523-536.

Bondi, M. W., Houston, W. S., Eyler, L. T., & Brown, G. G. (2005). fMRI evidence of compensatory mechanisms in older adults at genetic risk for Alzheimer disease. *Neurology*, 64(3), 501-508.

Bookheimer, S. Y., Strojwas, M. H., Cohen, M. S., Saunders, A. M., Pericak-Vance, M. A., Mazziotta, J. C., & Small, G. W. (2000). Patterns of brain activation in people at risk for Alzheimer's disease. *N Engl J Med*, 343(7), 450-456.

Borenstein Graves, A., Mortimer, J. A., Bowen, J. D., McCormick, W. C., McCurry, S. M., Schellenberg, G. D., & Larson, E. B. (2001). Head circumference and incident Alzheimer's disease: modification by apolipoprotein E. *Neurology*, 57(8), 1453-1460.

Braak, H., Alafuzoff, I., Arzberger, T., Kretzschmar, H., & Del Tredici, K. (2006). Staging of Alzheimer disease-associated neurofibrillary pathology using paraffin sections and immunocytochemistry. *Acta Neuropathol (Berl)*, 112(4), 389-404.

Braak, H., & Braak, E. (1998). Alzheimer's disease starts in early adulthood. *Neurobiology of Aging*, 19, S78.

Braak, H., Braak, E., & Bohl, J. (1993). Staging of Alzheimer-related cortical destruction. *Eur Neurol*, 33(6), 403-408.

Brazier, M. A. B., & Petsche, H. (Eds.). (1978). *Architectonics of the cerebral cortex.* New York: Raven Press.

Brun, A., Liu, X., & Erikson, C. (1995). Synapse loss and gliosis in the molecular layer of the cerebral cortex in Alzheimer's disease and in frontal lobe degeneration. *Neurodegeneration*, 4(2), 171-177.

Buonomano, D. V., & Merzenich, M. M. (1998). Cortical plasticity: from synapses to maps. *Annu Rev Neurosci*, 21, 149-186.

Burke, W. J., Miller, J. P., Rubin, E. H., Morris, J. C., Coben, L. A., Duchek, J., Wittels, I. G., & Berg, L. (1988). Reliability of the Washington University Clinical Dementia Rating. *Arch Neurol*, 45(1), 31-32.

Butefisch, C. M. (2006). Neurobiological bases of rehabilitation. *Neurol Sci*, 27 Suppl 1, S18-23.

Butler, A. J., & Wolf, S. L. (2007). Putting the brain on the map: use of transcranial magnetic stimulation to assess and induce cortical plasticity of upper-extremity movement. *Phys Ther*, 87(6), 719-736.

Castel-Lacanal, E., Gerdelat-Mas, A., Marque, P., Loubinoux, I., & Simonetta-Moreau, M. (2007). Induction of cortical plastic changes in wrist muscles by paired associative stimulation in healthy subjects and post-stroke patients. *Exp Brain Res*, 180(1), 113-122.

Chen, M. S., Huber, A. B., van der Haar, M. E., Frank, M., Schnell, L., Spillmann, A. A., Christ, F., & Schwab, M. E. (2000). Nogo-A is a myelin-associated neurite outgrowth inhibitor and an antigen for monoclonal antibody IN-1. *Nature*, 403(6768), 434-439.

Chenn, A., & Walsh, C. A. (2002). Regulation of cerebral cortical size by control of cell cycle exit in neural precursors. *Science*, 297(5580), 365-369.

Chenn, A., & Walsh, C. A. (2003). Increased neuronal production, enlarged forebrains and cytoarchitectural distortions in beta-catenin overexpressing transgenic mice. *Cereb Cortex*, 13(6), 599-606.

Chetelat, G., Landeau, B., Eustache, F., Mezenge, F., Viader, F., de la Sayette, V., Desgranges, B., & Baron, J. C. (2005). Using voxel-based morphometry to map the structural changes associated with rapid conversion in MCI: a longitudinal MRI study. *Neuroimage*, 27(4), 934-946.

Coffey, C. E., Saxton, J. A., Ratcliff, G., Bryan, R. N., & Lucke, J. F. (1999). Relation of education to brain size in normal aging: implications for the reserve hypothesis. *Neurology*, 53(1), 189-196.

Coffey, C. E., Wilkinson, W. E., Parashos, I. A., Soady, S. A., Sullivan, R. J., Patterson, L. J., Figiel, G. S., Webb, M. C., Spritzer, C. E., & Djang, W. T. (1992). Quantitative cerebral anatomy of the aging human brain: a cross-sectional study using magnetic resonance imaging. *Neurology*, 42(3 Pt 1), 527-536.

Coleman, P., Federoff, H., & Kurlan, R. (2004). A focus on the synapse for neuroprotection in Alzheimer disease and other dementias. *Neurology*, 63(7), 1155-1162.

Coleman, P. D., & Flood, D. G. (1987). Neuron numbers and dendritic extent in normal aging and Alzheimer's disease. *Neurobiol Aging*, 8(6), 521-545.

Conforto, A. B., Cohen, L. G., dos Santos, R. L., Scaff, M., & Marie, S. K. (2007). Effects of somatosensory stimulation on motor function in chronic cortico-subcortical strokes. *J Neurol*, 254(3), 333-339.

Consensus recommendations for the postmortem diagnosis of Alzheimer's disease. The National Institute on Aging, and Reagan Institute Working Group on Diagnostic Criteria for the Neuropathological Assessment of Alzheimer's Disease. (1997). *Neurobiol Aging*, 18(4 Suppl), S1-2.

Convit, A., de Asis, J., de Leon, M. J., Tarshish, C. Y., De Santi, S., & Rusinek, H. (2000). Atrophy of the medial occipitotemporal, inferior, and middle temporal gyri in non-demented elderly predict decline to Alzheimer's disease. *Neurobiol Aging*, 21(1), 19-26.

Costa-Mattioli, M., Gobert, D., Stern, E., Gamache, K., Colina, R., Cuello, C., Sossin, W., Kaufman, R., Pelletier, J., Rosenblum, K., Krnjevic, K., Lacaille, J. C., Nader, K., & Sonenberg, N. (2007). eIF2alpha phosphorylation bidirectionally regulates the switch from short- to long-term synaptic plasticity and memory. *Cell*, 129(1), 195-206.

Courchesne, E., Chisum, H. J., Townsend, J., Cowles, A., Covington, J., Egaas, B., Harwood, M., Hinds, S., & Press, G. A. (2000). Normal brain development and aging: quantitative analysis at in vivo MR imaging in healthy volunteers. *Radiology*, 216(3), 672-682.

Cracchiolo, J. R., Mori, T., Nazian, S. J., Tan, J., Potter, H., & Arendash, G. W. (2007). Enhanced cognitive activity-over and above social or physical activity-is required to protect Alzheimer's mice against cognitive impairment, reduce Abeta deposition, and increase synaptic immunoreactivity. *Neurobiol Learn Mem*.

Crystal, H., Dickson, D., Sliwinski, M., Masur, D., Blau, A., & Lipton, R. B. (1996). Associations of status and change measures of neuropsychological function with pathologic changes in elderly, originally nondemented subjects. *Arch Neurol*, 53, 82-87.

Dani, S. U., Pittella, J. E., Boehme, A., Hori, A., & Schneider, B. (1997). Progressive formation of neuritic plaques and neurofibrillary tangles is exponentially related to age and neuronal size. A morphometric study of three geographically distinct series of aging people. *Dement Geriatr Cogn Disord*, 8(4), 217-227.

Davis, D. G., Schmitt, F. A., Wekstein, D. R., & Markesbery, W. R. (1999). Alzheimer neuropathologic alterations in aged cognitively normal subjects. *J Neuropathol Exp Neurol*, 58(4), 376-388.

DeCarli, C., Massaro, J., Harvey, D., Hald, J., Tullberg, M., Au, R., Beiser, A., D'Agostino, R., & Wolf, P. A. (2005). Measures of brain morphology and infarction in the framingham heart study: establishing what is normal. *Neurobiol Aging*, 26(4), 491-510.

DeCarli, C., Murphy, D. G., Gillette, J. A., Haxby, J. V., Teichberg, D., Schapiro, M. B., & Horwitz, B. (1994). Lack of age-related differences in temporal lobe volume of very healthy adults. *AJNR Am.J Neuroradiol.*, 15, 689-696.

Deng, J., & Dunaevsky, A. (2005). Dynamics of dendritic spines and their afferent terminals: spines are more motile than presynaptic boutons. *Dev Bioll*, 277(2), 366-377.

Dickstein, D. L., Kabaso, D., Rocher, A. B., Luebke, J. I., Wearne, S. L., & Hof, P. R. (2007). Changes in the structural complexity of the aged brain. *Aging Cell*, 6(3), 275-284.

Disterhoft, J. F., & Oh, M. M. (2006). Learning, aging and intrinsic neuronal plasticity. *Trends Neurosci*, 29(10), 587-599.

Drapeau, E., Montaron, M. F., Aguerre, S., & Abrous, D. N. (2007). Learning-induced survival of new neurons depends on the cognitive status of aged rats. *J Neurosci*, 27(22), 6037-6044.

Driscoll, I., Resnick, S. M., Troncoso, J. C., An, Y., O'Brien, R., & Zonderman, A. B. (2006). Impact of Alzheimer's pathology on cognitive trajectories in nondemented elderly. *Ann Neurol*, 60(6), 688-695.

Elgh, E., Larsson, A., Eriksson, S., & Nyberg, L. (2003). Altered prefrontal brain activity in persons at risk for Alzheimer's disease: an fMRI study. *Int Psychogeriatr*, 15(2), 121-133.

Elias, M. F., Beiser, A., Wolf, P. A., Au, R., White, R. F., & D'Agostino, R. B. (2000). The preclinical phase of alzheimer disease: A 22-year prospective study of the Framingham Cohort. *Arch Neurol*, 57(6), 808-813.

Emerick, A. J., & Kartje, G. L. (2004). Behavioral recovery and anatomical plasticity in adult rats after cortical lesion and treatment with monoclonal antibody IN-1. *Behav Brain Res*, 152(2), 315-325.

Emerick, A. J., Neafsey, E. J., Schwab, M. E., & Kartje, G. L. (2003). Functional reorganization of the motor cortex in adult rats after cortical lesion and treatment with monoclonal antibody IN-1. *J Neurosci*, 23(12), 4826-4830.

Ertekin-Taner, N. (2007). Genetics of Alzheimer's disease: a centennial review. *Neurol Clin*, 25(3), 611-667.

Ethell, I. M., & Pasquale, E. B. (2005). Molecular mechanisms of dendritic spine development and remodeling. *Prog Neurobiol*, 75(3), 161-205.

Fagan, A. M., Roe, C. M., Xiong, C., Mintun, M. A., Morris, J. C., & Holtzman, D. M. (2007). Cerebrospinal fluid tau/beta-amyloid(42) ratio as a prediction of cognitive decline in nondemented older adults. *Arch Neurol*, 64(3), 343-349.

Fillenbaum, G. G., Peterson, B., & Morris, J. C. (1996). Estimating the validity of the clinical Dementia Rating Scale: the CERAD experience. Consortium to Establish a Registry for Alzheimer's Disease. *Aging (Milano)*, 8(6), 379-385.

Finehout, E. J., Franck, Z., Choe, L. H., Relkin, N., & Lee, K. H. (2007). Cerebrospinal fluid proteomic biomarkers for Alzheimer's disease. *Ann Neurol*, 61(2), 120-129.

Fleisher, A. S., Houston, W. S., Eyler, L. T., Frye, S., Jenkins, C., Thal, L. J., & Bondi, M. W. (2005). Identification of Alzheimer disease risk by functional magnetic resonance imaging. *Arch Neurol*, 62(12), 1881-1888.

Flicker, C., Ferris, S. H., & Reisberg, B. (1991). Mild cognitive impairment in the elderly: predictors of dementia. *Neurology*, 41(7), 1006-1009.

Flood, D. G., & Coleman, P. D. (1988). Neuron numbers and sizes in aging brain: comparisons of human, monkey, and rodent data. *Neurobiol Aging*, 9(5-6), 453-463.

Foster, T. C. (2007). Calcium homeostasis and modulation of synaptic plasticity in the aged brain. *Aging Cell*, 6(3), 319-325.

Freeman, S. H., Raju, S., Hyman, B. T., Frosch, M. P., & Irizarry, M. C. (2007). Plasma Abeta levels do not reflect brain Abeta levels. *J Neuropathol Exp Neurol*, 66(4), 264-271.

Friocourt, G., Poirier, K., Rakic, S., Parnavelas, J. G., & Chelly, J. (2006). The role of ARX in cortical development. *Eur J Neurosci*, 23(4), 869-876.

Galvin, J. E., Roe, C. M., Xiong, C., & Morris, J. C. (2006). Validity and reliability of the AD8 informant interview in dementia. *Neurology*, 67(11), 1942-1948.

Gauthier, S., Reisberg, B., Zaudig, M., Petersen, R. C., Ritchie, K., Broich, K., Belleville, S., Brodaty, H., Bennett, D., Chertkow, H., Cummings, J. L., de Leon, M., Feldman, H., Ganguli, M., Hampel, H., Scheltens, P., Tierney, M. C., Whitehouse, P., & Winblad, B. (2006). Mild cognitive impairment. *Lancet*, 367(9518), 1262-1270.

Geddes, J. W., Tekirian, T. L., Soultanian, N. S., Ashford, J. W., Davis, D. G., & Markesbery, W. R. (1997). Comparison of neuropathologic criteria for the diagnosis of Alzheimer's disease. *Neurobiol Aging*, 18(4 Suppl), S99-105.

Gervais, F., Paquette, J., Morissette, C., Krzywkowski, P., Yu, M., Azzi, M., Lacombe, D., Kong, X., Aman, A., Laurin, J., Szarek, W. A., & Tremblay, P. (2007). Targeting soluble Abeta peptide with Tramiprosate for the treatment of brain amyloidosis. *Neurobiol Aging*, 28(4), 537-547.

Giedd, J. N., Blumenthal, J., Jeffries, N. O., Rajapakse, J. C., Vaituzis, A. C., Liu, H., Berry, Y. C., Tobin, M., Nelson, J., & Castellanos, F. X. (1999). Development of the human corpus callosum during childhood and adolescence: a longitudinal MRI study. *Prog Neuropsychopharmacol Biol Psychiatry*, 23(4), 571-588.

Giedd, J.N., Blumenthal, J., Jeffries, N. O., Castellanos, F. X., Liu, H., Zijdenbos, A., Paus, T., Evans, A. C., Rapoport, J. L. (1999). Brain development during childhood and adolescence: a longitudinal MRI study. *Nat Neurosciencey*, 2(10), 861-863.

Gogtay, N., Giedd, J. N., Lusk, L., Hayashi, K. M., Greenstein, D., Vaituzis, A. C., Nugent, T. F., 3rd, Herman, D. H., Clasen, L. S., Toga, A. W., Rapoport, J. L., & Thompson, P. M. (2004). Dynamic mapping of human cortical development during childhood through early adulthood. *Proc Natl Acad Sci U S A*, 101(21), 8174-8179.

Golde, T. E. (2006). Disease modifying therapy for AD? *J Neurochem*, 99(3), 689-707.

Goldman, W. P., Price, J. L., Storandt, M., Grant, E. A., McKeel, D. W., Jr., Rubin, E. H., & Morris, J. C. (2001). Absence of cognitive impairment or decline in preclinical Alzheimer's disease. *Neurology*, 56(3), 361-367.

Gomes, R. A., Hampton, C., El-Sabeawy, F., Sabo, S. L., & McAllister, A. K. (2006). The dynamic distribution of TrkB receptors before, during, and after synapse formation between cortical neurons. *J Neurosci*, 26(44), 11487-11500.

Gongidi, V., Ring, C., Moody, M., Brekken, R., Sage, E. H., Rakic, P., & Anton, E. S. (2004). SPARC-like 1 regulates the terminal phase of radial glia-guided migration in the cerebral cortex. *Neuron*, 41(1), 57-69.

Good, C. D., Johnsrude, I. S., Ashburner, J., Henson, R. N., Friston, K. J., & Frackowiak, R. S. (2001). A voxel-based morphometric study of ageing in 465 normal adult human brains. *Neuroimage*, 14(1 Pt 1), 21-36.

Graff-Radford, N. R., Crook, J. E., Lucas, J., Boeve, B. F., Knopman, D. S., Ivnik, R. J., Smith, G. E., Younkin, L. H., Petersen, R. C., & Younkin, S. G. (2007). Association of low plasma Abeta42/Abeta40 ratios with increased imminent risk for mild cognitive impairment and Alzheimer disease. *Arch Neurol*, 64(3), 354-362.

Gralle, M., & Ferreira, S. T. (2007). Structure and functions of the human amyloid precursor protein: the whole is more than the sum of its parts. *Prog Neurobiol*, 82(1), 11-32.

Greenwood, P. M. (2000). The frontal aging hypothesis evaluated. *J Int Neuropsychol Soc*, 6(6), 705-726.

Grober, E., Lipton, R. B., Hall, C., & Crystal, H. (2000). Memory impairment on free and cued selective reminding predicts dementia. *Neurology*, 54(4), 827-832.

Guenette, S., Chang, Y., Hiesberger, T., Richardson, J. A., Eckman, C. B., Eckman, E. A., Hammer, R. E., & Herz, J. (2006). Essential roles for the FE65 amyloid precursor protein-interacting proteins in brain development. *Embo J*, 25(2), 420-431.

Head, D., Snyder, A. Z., Girton, L. E., Morris, J. C., & Buckner, R. L. (2005). Frontal-hippocampal double dissociation between normal aging and Alzheimer's disease. *Cereb Cortex*, 15(6), 732-739.

Head, E., Lott, I. T., Patterson, D., Doran, E., & Haier, R. J. (2007). Possible compensatory events in adult Down syndrome brain prior to the development of Alzheimer disease neuropathology: targets for nonpharmacological intervention. *J Alzheimers Dis*, 11(1), 61-76.

Ikonomovic, M. D., Uryu, K., Abrahamson, E. E., Ciallella, J. R., Trojanowski, J. Q., Lee, V. M., Clark, R. S., Marion, D. W., Wisniewski, S. R., & DeKosky, S. T. (2004). Alzheimer's pathology in human temporal cortex surgically excised after severe brain injury. *Exp Neurol*, 190(1), 192-203.

Israely, I., Costa, R. M., Xie, C. W., Silva, A. J., Kosik, K. S., & Liu, X. (2004). Deletion of the neuron-specific protein delta-catenin leads to severe cognitive and synaptic dysfunction. *Curr Biol*, 14(18), 1657-1663.

Jack, C. R., Jr., Petersen, R. C., Xu, Y., O'Brien, P. C., Smith, G. E., Ivnik, R. J., Boeve, B. F., Tangalos, E. G., & Kokmen, E. (2000). Rates of hippocampal atrophy correlate with change in clinical status in aging and AD. *Neurology*, 55(4), 484-489.

Jack, C. R., Jr., Petersen, R. C., Xu, Y. C., O'Brien, P. C., Smith, G. E., Ivnik, R. J., Boeve, B. F., Waring, S. C., Tangalos, E. G., & Kokmen, E. (1999). Prediction of AD with MRI-based hippocampal volume in mild cognitive impairment. *Neurology*, 52(7), 1397-1403.

Jack, C. R., Jr., Shiung, M. M., Gunter, J. L., O'Brien, P. C., Weigand, S. D., Knopman, D. S., Boeve, B. F., Ivnik, R. J., Smith, G. E., Cha, R. H., Tangalos, E. G., & Petersen, R. C. (2004). Comparison of different MRI brain atrophy rate measures with clinical disease progression in AD. *Neurology*, 62(4), 591-600.

Jack, C. R., Jr., Shiung, M. M., Weigand, S. D., O'Brien, P. C., Gunter, J. L., Boeve, B. F., Knopman, D. S., Smith, G. E., Ivnik, R. J., Tangalos, E. G., & Petersen, R. C. (2005). Brain atrophy rates predict subsequent clinical conversion in normal elderly and amnestic MCI. *Neurology*, 65(8), 1227-1231.

Jagust, W., Gitcho, A., Sun, F., Kuczynski, B., Mungas, D., & Haan, M. (2006). Brain imaging evidence of preclinical Alzheimer's disease in normal aging. *Ann Neurol*, 59(4), 673-681.

Jellinger, K. A. (1998). The neuropathological diagnosis of Alzheimer disease. *J Neural Transm Suppl*, 53, 97-118.

Jellinger, K. A., & Bancher, C. (1998). Neuropathology of Alzheimer's disease: a critical update. *J Neural Transm Suppl*, 54, 77-95.

Jernigan T.L., & Tallal, P. (1990). Late childhood changes in brain morphology observable with MRI. *Dev Med Child Neurol*, 32(5), 379-385.

Jernigan T,L., Trauner, D. A., Hesselink, J. R., & Tallal, P. A. (1991) Maturation of human cerebrum observed in vivo during adolescence. *Brain* 114, 2037-2049.

Jernigan, T. L., & Gamst, A. C. (2005). Changes in volume with age--consistency and interpretation of observed effects. *Neurobiol Aging*, 26(9), 1271-1274; discussion 1275-1278.

Johnson, K. A., Lopera, F., Jones, K., Becker, A., Sperling, R., Hilson, J., Londono, J., Siegert, I., Arcos, M., Moreno, S., Madrigal, L., Ossa, J., Pineda, N., Ardila, A., Roselli, M., Albert, M. S., Kosik, K. S., & Rios, A. (2001). Presenilin-1-associated abnormalities in regional cerebral perfusion. *Neurology*, 56(11), 1545-1551.

Karas, G. B., Scheltens, P., Rombouts, S. A., Visser, P. J., van Schijndel, R. A., Fox, N. C., & Barkhof, F. (2004). Global and local gray matter loss in mild cognitive impairment and Alzheimer's disease. *Neuroimage*, 23(2), 708-716.

Karunanayaka, P. R., Holland, S. K., Schmithorst, V. J., Solodkin, A., Chen, E. E., Szaflarski, J. P., & Plante, E. (2007). Age-related connectivity changes in fMRI data from children listening to stories. *Neuroimage*, 34(1), 349-360.

Kato, M., Das, S., Petras, K., Kitamura, K., Morohashi, K., Abuelo, D. N., Barr, M., Bonneau, D., Brady, A. F., Carpenter, N. J., Cipero, K. L., Frisone, F., Fukuda, T., Guerrini, R., Iida, E., Itoh, M., Lewanda, A. F., Nanba, Y., Oka, A., Proud, V. K., Saugier-Veber, P., Schelley, S. L., Selicorni, A., Shaner, R., Silengo, M., Stewart, F., Sugiyama, N., Toyama, J., Toutain, A., Vargas, A. L., Yanazawa, M., Zackai, E. H., & Dobyns, W. B. (2004). Mutations of ARX are associated with striking pleiotropy and consistent genotype-phenotype correlation. *Hum Mutat*, 23(2), 147-159.

Katzman, R. (1993). Education and the prevalence of dementia and Alzheimer's disease. *Neurology*, 43, 13-20.

Kaye, J. A., Swihart, T., Howieson, D., Dame, A., Moore, M. M., Karnos, T., Camicioli, R., Ball, M., Oken, B., & Sexton, G. (1997). Volume loss of the hippocampus and temporal lobe in healthy elderly persons destined to develop dementia. *Neurology*, 48(5), 1297-1304.

Kelley, B. J., & Petersen, R. C. (2007). Alzheimer's disease and mild cognitive impairment. *Neurol Clin*, 25(3), 577-609.

Kidron, D., Black, S. E., Stanchev, P., Buck, B., Szalai, J. P., Parker, J., Szekely, C., & Bronskill, M. J. (1997). Quantitative MR volumetry in Alzheimer's disease. Topographic markers and the effects of sex and education. *Neurology*, 49, 1504-1512.

Komuro, H., & Rakic, P. (1996). Intracellular Ca2+ fluctuations modulate the rate of neuronal migration. *Neuron*, 17(2), 275-285.

Komuro, H., & Rakic, P. (1998). Orchestration of neuronal migration by activity of ion channels, neurotransmitter receptors, and intracellular Ca2+ fluctuations. *J Neurobiol*, 37(1), 110-130.

Kovari, E., Gold, G., Herrmann, F. R., Canuto, A., Hof, P. R., Bouras, C., & Giannakopoulos, P. (2007). Cortical microinfarcts and demyelination affect cognition in cases at high risk for dementia. *Neurology*, 68(12), 927-931.

Kramer, A. F., & Erickson, K. I. (2007). Capitalizing on cortical plasticity: influence of physical activity on cognition and brain function. *Trends Cogn Sci*, 11(8), 342-348.

Kukar, T., Prescott, S., Eriksen, J. L., Holloway, V., Murphy, M. P., Koo, E. H., Golde, T. E., & Nicolle, M. M. (2007). Chronic administration of R-flurbiprofen attenuates learning impairments in transgenic amyloid precursor protein mice. *BMC Neurosci*, 8, 54.

Lemaitre, H., Crivello, F., Grassiot, B., Alperovitch, A., Tzourio, C., & Mazoyer, B. (2005). Age- and sex-related effects on the neuroanatomy of healthy elderly. *Neuroimage*, 26(3), 900-911.

Lessoway, V.A., Schulzer, M.,Wittmann, B. K., Gagnon, F. A., Wilson, R. D. (2008) Ultrasound fetal biometry charts for a North American Caucasian population. *J Clin Ultrasound* 26(9), 433-453.

Letenneur, L., Gilleron, V., Commenges, D., Helmer, C., Orgogozo, J. M., & Dartigues, J. F. (1999). Are sex and educational level independent predictors of dementia and Alzheimer's disease? Incidence data from the PAQUID project. *J Neurol Neurosurg Psychiatry*, 66(2), 177-183.

Liao, Y. C., Liu, R. S., Teng, E. L., Lee, Y. C., Wang, P. N., Lin, K. N., Chung, C. P., & Liu, H. C. (2005). Cognitive reserve: a SPECT study of 132 Alzheimer's disease patients with an education range of 0-19 years. *Dement Geriatr Cogn Disord*, 20(1), 8-14.

Lim, K. O., Zipursky, R. B., Watts, M. C., & Pfefferbaum, A. (1992). Decreased gray matter in normal aging: an in vivo magnetic resonance study. *J Gerontol*, 47(1), B26-30.

Lippman, J., & Dunaevsky, A. (2005). Dendritic spine morphogenesis and plasticity. *J Neurobiol*, 64(1), 47-57.

Liu, R. S., Lemieux, L., Bell, G. S., Sisodiya, S. M., Shorvon, S. D., Sander, J. W., & Duncan, J. S. (2003). A longitudinal study of brain morphometrics using quantitative magnetic resonance imaging and difference image analysis. *Neuroimage*, 20(1), 22-33.

Lockhart, A., Lamb, J. R., Osredkar, T., Sue, L. I., Joyce, J. N., Ye, L., Libri, V., Leppert, D., & Beach, T. G. (2007). PIB is a non-specific imaging marker of amyloid-beta (A{beta}) peptide-related cerebral amyloidosis. *Brain*.

Lonze, B. E., & Ginty, D. D. (2002). Function and regulation of CREB family transcription factors in the nervous system. *Neuron*, 35(4), 605-623.

Lorent, K., Overbergh, L., Moechars, D., De Strooper, B., Van Leuven, F., & Van den Berghe, H. (1995). Expression in mouse embryos and in adult mouse brain of three members of the amyloid precursor protein family, of the alpha-2-macroglobulin receptor/low density lipoprotein receptor-related protein and of its ligands apolipoprotein E, lipoprotein lipase, alpha-2-macroglobulin and the 40,000 molecular weight receptor-associated protein. *Neuroscience*, 65(4), 1009-1025.

Lovell, M. A., Xie, C., & Markesbery, W. R. (1998). Decreased glutathione transferase activity in brain and ventricular fluid in Alzheimer's disease. *Neurology*, 51(6), 1562-1566.

Mahncke, H. W., Connor, B. B., Appelman, J., Ahsanuddin, O. N., Hardy, J. L., Wood, R. A., Joyce, N. M., Boniske, T., Atkins, S. M., & Merzenich, M. M. (2006). Memory enhancement in healthy older adults using a brain plasticity-based training program: a randomized, controlled study. *Proc Natl Acad Sci U S A*, 103(33), 12523-12528.

Malaterre, J., Ramsay, R. G., & Mantamadiotis, T. (2007). Wnt-Frizzled signalling and the many paths to neural development and adult brain homeostasis. *Front Biosci*, 12, 492-506.

Mally, J., & Stone, T. W. (2007). New advances in the rehabilitation of CNS diseases applying rTMS. *Expert Rev Neurother*, 7(2), 165-177.

Marjaux, E., Hartmann, D., & De Strooper, B. (2004). Presenilins in memory, Alzheimer's disease, and therapy. *Neuron*, 42(2), 189-192.

Mark, V. W., Taub, E., & Morris, D. M. (2006). Neuroplasticity and constraint-induced movement therapy. *Eura Medicophys*, 42(3), 269-284.

Markesbery, W. R., Schmitt, F. A., Kryscio, R. J., Davis, D. G., Smith, C. D., & Wekstein, D. R. (2006). Neuropathologic substrate of mild cognitive impairment. *Arch Neurol*, 63(1), 38-46.

Markus, T. M., Tsai, S. Y., Bollnow, M. R., Farrer, R. G., O'Brien, T. E., Kindler-Baumann, D. R., Rausch, M., Rudin, M., Wiessner, C., Mir, A. K., Schwab, M. E., & Kartje, G. L. (2005). Recovery and brain reorganization after stroke in adult and aged rats. *Ann Neurol*, 58(6), 950-953.

Marui, W., Iseki, E., Kato, M., Akatsu, H., & Kosaka, K. (2004). Pathological entity of dementia with Lewy bodies and its differentiation from Alzheimer's disease. *Acta Neuropathol (Berl)*, 108(2), 121-128.

Masliah, E., Terry, R. D., DeTeresa, R. M., & Hansen, L. A. (1989). Immunohistochemical quantification of the synapse-related protein synaptophysin in Alzheimer disease. *Neurosci Lett*, 103(2), 234-239.

Masur, D. M., Sliwinski, M., Lipton, R. B., Blau, A. D., & Crystal, H. A. (1994). Neuropsychological prediction of dementia and the absence of dementia in healthy elderly persons. *Neurology*, 44(8), 1427-1432.

Mattson, M. P. (2007). Calcium and neurodegeneration. *Aging Cell*, 6(3), 337-350.

McDowell, I., Xi, G., Lindsay, J., & Tierney, M. (2007). Mapping the connections between education and dementia. *J Clin Exp Neuropsychol*, 29(2), 127-141.

McKhann, G., Drachman, D., Folstein, M., Katzman, R., Price, D., & Stadlan, E. M. (1984). Clinical diagnosis of Alzheimer's disease: report of the NINCDS-ADRDA Work Group under the auspices of Department of Health and Human Services Task Force on Alzheimer's Disease. *Neurology*, 34(7), 939-944.

Metzler-Baddeley, C. (2007). A review of cognitive impairments in dementia with Lewy bodies relative to Alzheimer's disease and Parkinson's disease with dementia. *Cortex*, 43(5), 583-600.

Mileusnic, R., Lancashire, C. L., & Rose, S. P. (2005). Amyloid precursor protein: from synaptic plasticity to Alzheimer's disease. *Ann N Y Acad Sci*, 1048, 149-165.

Miyasaka, T., Watanabe, A., Saito, Y., Murayama, S., Mann, D. M., Yamazaki, M., Ravid, R., Morishima-Kawashima, M., Nagashima, K., & Ihara, Y. (2005). Visualization of newly deposited tau in neurofibrillary tangles and neuropil threads. *J Neuropathol Exp Neurol*, 64(8), 665-674.

Morris, J. C. (1997). Clinical dementia rating: a reliable and valid diagnostic and staging measure for dementia of the Alzheimer type. *Int Psychogeriatr*, 9 Suppl 1, 173-176; discussion 177-178.

Morris, J. C., McKeel, D. W., Jr., Storandt, M., Rubin, E. H., Price, J. L., Grant, E. A., Ball, M. J., & Berg, L. (1991). Very mild Alzheimer's disease: informant-based clinical, psychometric, and pathologic distinction from normal aging. *Neurology*, 41(4), 469-478.

Morris, J. C., & Price, A. L. (2001). Pathologic correlates of nondemented aging, mild cognitive impairment, and early-stage Alzheimer's disease. *J Mol Neurosci*, 17(2), 101-118.

Mortimer, J. A. (1997). Brain reserve and the clinical expression of Alzheimer's disease. *Geriatrics*, 52 Suppl 2, S50-53.

Mortimer, J. A., Borenstein, A. R., Gosche, K. M., & Snowdon, D. A. (2005). Very early detection of Alzheimer neuropathology and the role of brain reserve in modifying its clinical expression. *J Geriatr Psychiatry Neurol*, 18(4), 218-223.

Mosconi, L., Herholz, K., Prohovnik, I., Nacmias, B., De Cristofaro, M. T., Fayyaz, M., Bracco, L., Sorbi, S., & Pupi, A. (2005). Metabolic interaction between ApoE genotype and onset age in Alzheimer's disease: implications for brain reserve. *J Neurol Neurosurg Psychiatry*, 76(1), 15-23.

Mothet, J. P., Rouaud, E., Sinet, P. M., Potier, B., Jouvenceau, A., Dutar, P., Videau, C., Epelbaum, J., & Billard, J. M. (2006). A critical role for the glial-derived neuromodulator D-serine in the age-related deficits of cellular mechanisms of learning and memory. *Aging Cell*, 5(3), 267-274.

Mueller, E. A., Moore, M. M., Kerr, D. C., Sexton, G., Camicioli, R. M., Howieson, D. B., Quinn, J. F., & Kaye, J. A. (1998). Brain volume preserved in healthy elderly through the eleventh decade. *Neurology*, 51(6), 1555-1562.

Nagata, K., Basugi, N., Fukushima, T., Tango, T., Suzuki, I., Kaminuma, T., & Kurashina, S. (1987). A quantitative study of physiological cerebral atrophy with aging. A statistical analysis of the normal range. *Neuroradiology*, 29(4), 327-332.

Newell, K. L., Hyman, B. T., Growdon, J. H., & Hedley-Whyte, E. T. (1999). Application of the National Institute on Aging (NIA)-Reagan Institute criteria for the neuropathological diagnosis of Alzheimer disease. *J Neuropathol Exp Neurol*, 58(11), 1147-1155.

Noctor, S,C., Martinez-Cerdeno, V., & Kriegstein, A. R. (2007). Contribution of intermediate progenitor cells to cortical histogenesis. *Arch Neurol* 64(5), 639-642.

Nordberg, A. (2007). Amyloid imaging in Alzheimer's disease. *Curr Opin Neurol*, 20(4), 398-402.

Ohm, T. G., Muller, H., Braak, H., & Bohl, J. (1995). Close-meshed prevalence rates of different stages as a tool to uncover the rate of Alzheimer's disease-related neurofibrillary changes. *Neuroscience*, 64(1), 209-217.

Olsson, A., Csajbok, L., Ost, M., Hoglund, K., Nylen, K., Rosengren, L., Nellgard, B., & Blennow, K. (2004). Marked increase of beta-amyloid(1-42) and amyloid precursor protein in ventricular cerebrospinal fluid after severe traumatic brain injury. *J Neurol*, 251(7), 870-876.

Papadopoulos, C. M., Tsai, S. Y., Alsbiei, T., O'Brien, T. E., Schwab, M. E., & Kartje, G. L. (2002). Functional recovery and neuroanatomical plasticity following middle cerebral artery occlusion and IN-1 antibody treatment in the adult rat. *Ann Neurol*, 51(4), 433-441.

Pennanen, C., Testa, C., Laakso, M. P., Hallikainen, M., Helkala, E. L., Hanninen, T., Kivipelto, M., Kononen, M., Nissinen, A., Tervo, S., Vanhanen, M., Vanninen, R., Frisoni, G. B., & Soininen, H. (2005). A voxel based morphometry study on mild cognitive impairment. *J Neurol Neurosurg Psychiatry*, 76(1), 11-14.

Perneczky, R., Diehl-Schmid, J., Drzezga, A., & Kurz, A. (2007). Brain reserve capacity in frontotemporal dementia: a voxel-based (18)F-FDG PET study. *Eur J Nucl Med Mol Imaging*, 34(7), 1082-1087.

Perneczky, R., Diehl-Schmid, J., Pohl, C., Drzezga, A., & Kurz, A. (2007). Non-fluent progressive aphasia: cerebral metabolic patterns and brain reserve. *Brain Res*, 1133(1), 178-185.

Perneczky, R., Drzezga, A., Diehl-Schmid, J., Schmid, G., Wohlschlager, A., Kars, S., Grimmer, T., Wagenpfeil, S., Monsch, A., & Kurz, A. (2006). Schooling mediates *Brain Res*erve in Alzheimer's disease: findings of fluoro-deoxy-glucose-positron emission tomography. *J Neurol Neurosurg Psychiatry*, 77(9), 1060-1063.

Petersen, R. C. (1998). Clinical subtypes of Alzheimer's disease. *Dement Geriatr Cogn Disord*, 9 Suppl 3, 16-24.

Petersen, R. C., Parisi, J. E., Dickson, D. W., Johnson, K. A., Knopman, D. S., Boeve, B. F., Jicha, G. A., Ivnik, R. J., Smith, G. E., Tangalos, E. G., Braak, H., & Kokmen, E. (2006). Neuropathologic features of amnestic mild cognitive impairment. *Arch Neurol*, 63(5), 665-672.

Petersen, R. C., Smith, G. E., Waring, S. C., Ivnik, R. J., Tangalos, E. G., & Kokmen, E. (1999). Mild cognitive impairment: clinical characterization and outcome. *Arch Neurol*, 56(3), 303-308.

Pfefferbaum, A., Mathalon, D. H., Sullivan, E. V., Rawles, J. M., Zipursky, R. B., & Lim, K. O. (1994). A quantitative magnetic resonance imaging study of changes in brain morphology from infancy to late adulthood. *Arch Neurol*, 51(9), 874-887.

Pfefferbaum, A., Sullivan, E. V., Swan, G. E., & Carmelli, D. (2000). Brain structure in men remains highly heritable in the seventh and eighth decades of life. *Neurobiol Aging*, 21(1), 63-74.

Plassman, B. L., Welsh, K. A., Helms, M., Brandt, J., Page, W. F., & Breitner, J. C. S. (1995). Intelligence and education as predictors of cognitive state in late life: a 50-year follow-up. *Neurology*, 45, 1446-1450.

Potter, G. G., Plassman, B. L., Helms, M. J., Foster, S. M., & Edwards, N. W. (2006). Occupational characteristics and cognitive performance among elderly male twins. *Neurology*, 67(8), 1377-1382.

Price, D. L., Thinakaran, G., Borchelt, D. R., Martin, L. J., Crain, B. J., Sisodia, S. S., & Troncoso, J. C. (1998). Neuropathology of Alzheimer's disease and animal models. In W. R. Markesbery (Ed.), *Neuropathology of dementing disorders*. London: Arnold.

Rakic, P. (1990). Principles of neural cell migration. *Experientia*, 46(9), 882-891.

Rakic, P. (2007). The radial edifice of cortical architecture: From neuronal silhouettes to genetic engineering. *Brain Res Rev*.

Ramachandran, V. S. (2005a). Plasticity and functional recovery in neurology. *Clin Med*, 5(4), 368-373.

Ramachandran, V. S. (2005b). *Plasticity and functional recovery in neurology*. (1470-2118 (Print)).

Ramachandran, V. S., & Rogers-Ramachandran, D. (2000). Phantom limbs and neural plasticity. *Arch Neurol*, 57(3), 317-320.

Ramic, M., Emerick, A. J., Bollnow, M. R., O'Brien, T. E., Tsai, S. Y., & Kartje, G. L. (2006). Axonal plasticity is associated with motor recovery following amphetamine treatment combined with rehabilitation after brain injury in the adult rat. *Brain Res*, 1111(1), 176-186.

Rauch, R. A., & Jinkins, J. R. (1994). Analysis of cross-sectional area measurements of the corpus callosum adjusted for brain size in male and female subjects from childhood to adulthood. *Behav Brain Res*, 64(1-2), 65-78.

Raz, N., Gunning, F. M., Head, D., Dupuis, J. H., McQuain, J., Briggs, S. D., Loken, W. J., Thornton, A. E., & Acker, J. D. (1997). Selective aging of the human cerebral cortex observed in vivo: differential vulnerability of the prefrontal gray matter. *Cereb Cortex*, 7(3), 268-282.

Reiman, E. M., Caselli, R. J., Yun, L. S., Kewei, C., Bandy, D., Minishima, S., Thibideau, S. N., & Osborne, D. (1996). Preclinical evidence of Alzheimer's disease in persons homozygous for the e4 allele for apolipoprotein E. *NEJM*, 334, 752-758.

Reiman, E. M., Webster, J. A., Myers, A. J., Hardy, J., Dunckley, T., Zismann, V. L., Joshipura, K. D., Pearson, J. V., Hu-Lince, D., Huentelman, M. J., Craig, D. W., Coon, K. D., Liang, W. S., Herbert, R. H., Beach, T., Rohrer, K. C., Zhao, A. S., Leung, D., Bryden, L., Marlowe, L., Kaleem, M., Mastroeni, D., Grover, A., Heward, C. B., Ravid, R., Rogers, J., Hutton, M. L., Melquist, S., Petersen, R. C., Alexander, G. E., Caselli, R. J., Kukull, W., Papassotiropoulos, A., & Stephan, D. A. (2007). GAB2 alleles modify Alzheimer's risk in APOE epsilon4 carriers. *Neuron*, 54(5), 713-720.

Resnick, S. M., Pham, D. L., Kraut, M. A., Zonderman, A. B., & Davatzikos, C. (2003). Longitudinal magnetic resonance imaging studies of older adults: a shrinking brain. *J Neurosci*, 23(8), 3295-3301.

Roe, C. M., Xiong, C., Miller, J. P., & Morris, J. C. (2007). Education and Alzheimer disease without dementia: support for the cognitive reserve hypothesis. *Neurology*, 68(3), 223-228.

Rovio, S., Kareholt, I., Viitanen, M., Winblad, B., Tuomilehto, J., Soininen, H., Nissinen, A., & Kivipelto, M. (2007). Work-related physical activity and the risk of dementia and Alzheimer's disease. *Int J Geriatr Psychiatry*, 22(9), 874-882.

Rowe, C. C., Ng, S., Ackermann, U., Gong, S. J., Pike, K., Savage, G., Cowie, T. F., Dickinson, K. L., Maruff, P., Darby, D., Smith, C., Woodward, M., Merory, J., Tochon-Danguy, H., O'Keefe, G., Klunk, W. E., Mathis, C. A., Price, J. C., Masters, C. L., & Villemagne, V. L. (2007). Imaging beta-amyloid burden in aging and dementia. *Neurology*, 68(20), 1718-1725.

Salat, D. H., Kaye, J. A., & Janowsky, J. S. (1999). Prefrontal gray and white matter volumes in healthy aging and Alzheimer disease. *Arch Neurol*, 56(3), 338-344.

Sarkisian, M. R., Bartley, C. M., Chi, H., Nakamura, F., Hashimoto-Torii, K., Torii, M., Flavell, R. A., & Rakic, P. (2006). MEKK4 signaling regulates filamin expression and neuronal migration. *Neuron*, 52(5), 789-801.

Saura, C. A., Choi, S. Y., Beglopoulos, V., Malkani, S., Zhang, D., Shankaranarayana Rao, B. S., Chattarji, S., Kelleher, R. J., 3rd, Kandel, E. R., Duff, K., Kirkwood, A., & Shen, J. (2004). Loss of presenilin function causes impairments of memory and synaptic plasticity followed by age-dependent neurodegeneration. *Neuron*, 42(1), 23-36.

Scahill, R. I., Frost, C., Jenkins, R., Whitwell, J. L., Rossor, M. N., & Fox, N. C. (2003). A longitudinal study of brain volume changes in normal aging using serial registered magnetic resonance imaging. *Arch Neurol*, 60(7), 989-994.

Scarmeas, N., Zarahn, E., Anderson, K. E., Honig, L. S., Park, A., Hilton, J., Flynn, J., Sackeim, H. A., & Stern, Y. (2004). Cognitive reserve-mediated modulation of positron emission tomographic activations during memory tasks in Alzheimer disease. *Arch Neurol*, 61(1), 73-78.

Scheff, S. W., & Price, D. A. (2006). Alzheimer's disease-related alterations in synaptic density: neocortex and hippocampus. *J Alzheimers Dis*, 9(3 Suppl), 101-115.

Scheff, S. W., Price, D. A., Schmitt, F. A., & Mufson, E. J. (2006). Hippocampal synaptic loss in early Alzheimer's disease and mild cognitive impairment. *Neurobiol Aging*, 27(10), 1372-1384.

Scherder, E., Eggermont, L., Sergeant, J., & Boersma, F. (2007). Physical activity and cognition in Alzheimer's disease: relationship to vascular risk factors, executive functions and gait. *Rev Neurosci*, 18(2), 149-158.

Schmidt, C., Lepsverdize, E., Chi, S. L., Das, A. M., Pizzo, S. V., Dityatev, A., & Schachner, M. (2007). Amyloid precursor protein and amyloid beta-peptide bind to ATP synthase and regulate its activity at the surface of neural cells. *Mol Psychiatry*.

Schmithorst, V. J., Holland, S. K., & Dardzinski, B. J. (2007). Developmental differences in white matter architecture between boys and girls. *Hum Brain Mapp*.

Schmithorst, V. J., Holland, S. K., & Plante, E. (2006). Object identification and lexical/semantic access in children: A functional magnetic resonance imaging study of word-picture matching. *Hum Brain Mapp*.

Schmithorst, V. J., Wilke, M., Dardzinski, B. J., & Holland, S. K. (2002). Correlation of white matter diffusivity and anisotropy with age during childhood and adolescence: a cross-sectional diffusion-tensor MR imaging study. *Radiology*, 222(1), 212-218.

Schmitt, F. A., Davis, D. G., Wekstein, D. R., Smith, C. D., Ashford, J. W., & Markesbery, W. R. (2000). "Preclinical" AD revisited: neuropathology of cognitively normal older adults. *Neurology*, 55(3), 370-376.

Schneider, J. A., Arvanitakis, Z., Bang, W., & Bennett, D. A. (2007). Mixed brain pathologies account for most dementia cases in community-dwelling older persons. *Neurology*.

Schofield, P. W., Mosesson, R. E., Stern, Y., & Mayeux, R. (1995). The age at onset of Alzheimer's disease and an intracranial area measurement. A relationship. *Arch.Neurol.*, 52, 95-98.

Sharma, K., Fong, D. K., & Craig, A. M. (2006). Postsynaptic protein mobility in dendritic spines: long-term regulation by synaptic NMDA receptor activation. *Mol Cell Neurosci*, 31(4), 702-712.

Sidman, R. L., & Rakic, P. (1982). Development of the human central nervous system. In W. Haymaker & R. D. Adams (Eds.), *Histology and Histopathology of the Nervous System* (pp. 1 -112). Springfield Ill.: Thomas.

Silva, A. J., & Giese, K. P. (1994). Plastic genes are in! *Curr Opin Neurobiol*, 4(3), 413-420.

Small, D. H. (2004). Mechanisms of synaptic homeostasis in Alzheimer's disease. *Curr Alzheimer Res*, 1(1), 27-32.

Small, G. W., Ercoli, L. M., Silverman, D. H., Huang, S. C., Komo, S., Bookheimer, S. Y., Lavretsky, H., Miller, K., Siddarth, P., Rasgon, N. L., Mazziotta, J. C., Saxena, S., Wu, H. M., Mega, M. S., Cummings, J. L., Saunders, A. M., Pericak-Vance, M. A., Roses, A. D., Barrio, J. R., & Phelps, M. E. (2000). Cerebral metabolic and cognitive decline in persons at genetic risk for Alzheimer's disease. *Proc Natl Acad Sci U S A*, 97(11), 6037-6042.

Small, S. A., Stern, Y., Tang, M., & Mayeux, R. (1999). Selective decline in memory function among healthy elderly. *Neurology*, 52(7), 1392-1396.

Smith, C. D., Andersen, A. H., Kryscio, R. J., Schmitt, F. A., Blonder, L. X., Kindy, M. S., & Avison, M. J. (1999). Altered brain activation in normal subjects at risk for Alzheimer's disease. *Neurology*, 53, 1391-1396.

Smith, C. D., Andersen, A. H., Kryscio, R. J., Schmitt, F. A., Kindy, M. S., Blonder, L. X., & Avison, M. J. (1999). Altered brain activation in cognitively intact individuals at high risk for Alzheimer's disease. *Neurology*, 53(7), 1391-1396.

Smith, C. D., Andersen, A. H., Kryscio, R. J., Schmitt, F. A., Kindy, M. S., Blonder, L. X., & Avison, M. J. (2002). Women at risk for AD show increased parietal activation during a fluency task. *Neurology*, 58(8), 1197-1202.

Smith, C. D., Chebrolu, H., Wekstein, D. R., Schmitt, F. A., Jicha, G. A., Cooper, G., & Markesbery, W. R. (2007). Brain structural alterations before mild cognitive impairment. *Neurology*, 68(16), 1268-1273.

Smith, C. D., Chebrolu, H., Wekstein, D. R., Schmitt, F. A., & Markesbery, W. R. (2007). Age and gender effects on human brain anatomy: a voxel-based morphometric study in healthy elderly. *Neurobiol Aging*, 28(7), 1075-1087.

Smith, C. D., Kryscio, R. J., Schmitt, F. A., Lovell, M. A., Blonder, L. X., Rayens, W. S., & Andersen, A. H. (2005). Longitudinal functional alterations in asymptomatic women at risk for Alzheimer's disease. *J Neuroimaging*, 15(3), 271-277.

Smith, D. E., Roberts, J., Gage, F. H., & Tuszynski, M. H. (1999). Age-associated neuronal atrophy occurs in the primate brain and is reversible by growth factor gene therapy. *Proc Natl Acad Sci U S A*, 96(19), 10893-10898.

Smyth, K. A., Fritsch, T., Cook, T. B., McClendon, M. J., Santillan, C. E., & Friedland, R. P. (2004). Worker functions and traits associated with occupations and the development of AD. *Neurology*, 63(3), 498-503.

Snowdon, D. A., Greiner, L. H., Mortimer, J. A., Riley, K. P., Greiner, P. A., & Markesbery, W. R. (1997). Brain infarction and the clinical expression of Alzheimer disease. The Nun Study. *Jama*, 277(10), 813-817.

Snowdon, D. A., Kemper, S. J., Mortimer, J. A., Greiner, L. H., Wekstein, D. R., & Markesbery, W. R. (1996). Linguistic ability in early life and cognitive function and Alzheimer's disease in late life. Findings from the Nun Study . *JAMA*, 275, 528-532.

Sowell, E.R., Thompson, P. M., Tessner, K. D., & Toga, A. W. (2001). Mapping continued brain growth and gray matter density reduction in dorsal frontal cortex: Inverse relationships during postadolescent brain maturation. *J Neurosci* 21(22), 8819-8829.

Sperling, R. (2007). Functional MRI studies of associative encoding in normal aging, mild cognitive impairment, and Alzheimer's disease. *Ann N Y Acad Sci*, 1097, 146-155.

Stefanova, E., Nilsson, A., Andersson, J., Axelman, K., Langstrom, B., Viitanen, M., Lannfelt, L., & Nordberg, A. (2002). Abnormalities in cerebral glucose metabolism of familial Alzheimer's disease associated with 670/671 APP mutation and presenilin 1 gene mutations. *Paper presented at the ICAD*, Kyoto, Japan.

Steiner, B., Wolf, S., & Kempermann, G. (2006). Adult neurogenesis and neurodegenerative disease. *Regen Med*, 1(1), 15-28.

Stern, R. A., Silva, S. G., Chaisson, N., & Evans, D. L. (1996). Influence of cognitive reserve on neuropsychological functioning in asymptomatic human immunodeficiency virus-1 infection. *Arch Neurol*, 53(2), 148-153.

Stern, Y. (2002). What is cognitive reserve? Theory and research application of the reserve concept. *J Int Neuropsychol Soc*, 8(3), 448-460.

Stern, Y., Alexander, G. E., Prohovnik, I., & Mayeux, R. (1992). Inverse relationship between education and parietotemporal perfusion deficit in Alzheimer's disease. *Ann Neurol*, 32(3), 371-375.

Stern, Y., Gurland, B., Tatemichi, T. K., Tang, M. X., Wilder, D., & Mayeux, R. (1994). Influence of education and occupation on the incidence of ALzheimer's disease. *JAMA*, 13, 1004-1010.

Stettler, D. D., Yamahachi, H., Li, W., Denk, W., & Gilbert, C. D. (2006). Axons and synaptic boutons are highly dynamic in adult visual cortex. *Neuron*, 49(6), 877-887.

Storandt, M., Grant, E. A., Miller, J. P., & Morris, J. C. (2006). Longitudinal course and neuropathologic outcomes in original vs revised MCI and in pre-MCI. *Neurology*, 67(3), 467-473.

Stromme, P., Mangelsdorf, M. E., Shaw, M. A., Lower, K. M., Lewis, S. M., Bruyere, H., Lutcherath, V., Gedeon, A. K., Wallace, R. H., Scheffer, I. E., Turner, G., Partington, M., Frints, S. G., Fryns, J. P., Sutherland, G. R., Mulley, J. C., & Gecz, J. (2002). Mutations in the human ortholog of Aristaless cause X-linked mental retardation and epilepsy. *Nat Genet*, 30(4), 441-445.

Sunderland, A., & Tuke, A. (2005). Neuroplasticity, learning and recovery after stroke: a critical evaluation of constraint-induced therapy. *Neuropsychol Rehabil*, 15(2), 81-96.

Sweatt, J. D. (2001). The neuronal MAP kinase cascade: a biochemical signal integration system subserving synaptic plasticity and memory. *J Neurochem*, 76(1), 1-10.

Sweatt, J. D., Weeber, E. J., & Lombroso, P. J. (2003). Genetics of childhood disorders: LI. Learning and memory, Part 4: Human cognitive disorders and the ras/ERK/CREB pathway. *J Am Acad Child Adolesc Psychiatry*, 42(6), 741-744.

Szaflarski, J. P., Holland, S. K., Schmithorst, V. J., & Byars, A. W. (2006). fMRI study of language lateralization in children and adults. *Hum Brain Mapp*, 27(3), 202-212.

Szczygielski, J., Mautes, A., Steudel, W. I., Falkai, P., Bayer, T. A., & Wirths, O. (2005). Traumatic brain injury: cause or risk of Alzheimer's disease? A review of experimental studies. *J Neural Transm*, 112(11), 1547-1564.

Tao-Cheng, J. H. (2006). Activity-related redistribution of presynaptic proteins at the active zone. *Neuroscience*, 141(3), 1217-1224.

Terry, R. D. (2006). Alzheimer's disease and the aging brain. *J Geriatr Psychiatry Neurol*, 19(3), 125-128.

Thomas, G. M., & Huganir, R. L. (2004). MAPK cascade signalling and synaptic plasticity. *Nat Rev Neurosci*, 5(3), 173-183.

Tierney, M. C., Yao, C., Kiss, A., & McDowell, I. (2005). Neuropsychological tests accurately predict incident Alzheimer disease after 5 and 10 years. *Neurology*, 64(11), 1853-1859.

Trivedi, M. A., Schmitz, T. W., Ries, M. L., Torgerson, B. M., Sager, M. A., Hermann, B. P., Asthana, S., & Johnson, S. C. (2006). Reduced hippocampal activation during episodic encoding in middle-aged individuals at genetic risk of Alzheimer's disease: a cross-sectional study. *BMC* Med, 4, 1.

Tsanov, M., & Manahan-Vaughan, D. (2007). The adult visual cortex expresses dynamic synaptic plasticity that is driven by the light/dark cycle. *J Neurosci*, 27(31), 8414-8421.

Turner, P. R., O'Connor, K., Tate, W. P., & Abraham, W. C. (2003). Roles of amyloid precursor protein and its fragments in regulating neural activity, plasticity and memory. *Prog Neurobiol*, 70(1), 1-32.

Unverzagt, F. W., Hui, S. L., Farlow, M. R., Hall, K. S., & Hendrie, H. C. (1998). Cognitive decline and education in mild dementia. *Neurology*, 50(1), 181-185.

Uylings, H. B., & de Brabander, J. M. (2002). Neuronal changes in normal human aging and Alzheimer's disease. *Brain Cogn*, 49(3), 268-276.

Van der Borght, K., Havekes, R., Bos, T., Eggen, B. J., & Van der Zee, E. A. (2007). Exercise improves memory acquisition and retrieval in the Y-maze task: relationship with hippocampal neurogenesis. *Behav Neurosci*, 121(2), 324-334.

Walhovd, K. B., Fjell, A. M., Reinvang, I., Lundervold, A., Dale, A. M., Eilertsen, D. E., Quinn, B. T., Salat, D., Makris, N., & Fischl, B. (2005). Effects of age on volumes of cortex, white matter and subcortical structures. *Neurobiol Aging*, 26(9), 1261-1270; discussion 1275-1268.

Walker-Batson, D., Curtis, S., Natarajan, R., Ford, J., Dronkers, N., Salmeron, E., Lai, J., & Unwin, D. H. (2001). A double-blind, placebo-controlled study of the use of amphetamine in the treatment of aphasia. *Stroke*, 32(9), 2093-2098.

Walsh, D. M., & Selkoe, D. J. (2004). Deciphering the molecular basis of memory failure in Alzheimer's disease. *Neuron*, 44(1), 181-193.

Waltereit, R., & Weller, M. (2003). Signaling from cAMP/PKA to MAPK and synaptic plasticity. *Mol Neurobiol*, 27(1), 99-106.

Weisman, D., Cho, M., Taylor, C., Adame, A., Thal, L. J., & Hansen, L. A. (2007). In dementia with Lewy bodies, Braak stage determines phenotype, not Lewy body distribution. *Neurology*, 69(4), 356-359.

Wilke, M., Krageloh-Mann, I., & Holland, S. K. (2007). Global and local development of gray and white matter volume in normal children and adolescents. *Exp Brain Res*, 178(3), 296-307.

Wilson, R. S., Bennett, D. A., Bienias, J. L., Aggarwal, N. T., Mendes De Leon, C. F., Morris, M. C., Schneider, J. A., & Evans, D. A. (2002). Cognitive activity and incident AD in a population-based sample of older persons. *Neurology*, 59(12), 1910-1914.

Wines-Samuelson, M., & Shen, J. (2005). Presenilins in the developing, adult, and aging cerebral cortex. *Neuroscientist*, 11(5), 441-451.

Wishart, H. A., Saykin, A. J., McAllister, T. W., Rabin, L. A., McDonald, B. C., Flashman, L. A., Roth, R. M., Mamourian, A. C., Tsongalis, G. J., & Rhodes, C. H. (2006). Regional brain atrophy in cognitively intact adults with a single APOE epsilon4 allele. *Neurology*, 67(7), 1221-1224.

Wishart, H. A., Saykin, A. J., Rabin, L. A., Santulli, R. B., Flashman, L. A., Guerin, S. J., Mamourian, A. C., Belloni, D. R., Rhodes, C. H., & McAllister, T. W. (2006). Increased brain activation during working memory in cognitively intact adults with the APOE epsilon4 allele. *Am J Psychiatry*, 163(9), 1603-1610.

Wisniewski, H. M., & Silverman, W. (1997). Diagnostic criteria for the neuropathological assessment of Alzheimer's disease: current status and major issues. *Neurobiol Aging*, 18(4 Suppl), S43-50.

Wolf, H., Julin, P., Gertz, H. J., Winblad, B., & Wahlund, L. O. (2004). Intracranial volume in mild cognitive impairment, Alzheimer's disease and vascular dementia: evidence for Brain Reserve? *Int J Geriatr Psychiatry*, 19(10), 995-1007.

Xu, Y., Jack, C. R., Jr., O'Brien, P. C., Kokmen, E., Smith, G. E., Ivnik, R. J., Boeve, B. F., Tangalos, R. G., & Petersen, R. C. (2000). Usefulness of MRI measures of entorhinal cortex versus hippocampus in AD. *Neurology*, 54(9), 1760-1767.

Y Cajal, S. R. (1995). *Histology of the nervous system* (S. Swanson & L. W. Swanson, Trans. Vol. 6). New York: Oxford University Press.

Yaffe, K., Petersen, R. C., Lindquist, K., Kramer, J., & Miller, B. (2006). Subtype of mild cognitive impairment and progression to dementia and death. *Dement Geriatr Cogn Disord*, 22(4), 312-319.

Yakovlev, P. I., & Lecours, A. R. (1967). The myelinogenetic cycles of regional maturation of the brain. In A. Mankowski (Ed.), *Regional development of the brain in early life* (pp. 3-69). Philadelphia: Davis.

Zacharia, A., Zimine, S., Lovblad, K. O., Warfield, S., Thoeny, H., Ozdoba, C., Bossi, E., Kreis, R., Boesch, C., Schroth, G., & Huppi, P. S. (2006). Early assessment of brain maturation by MR imaging segmentation in neonates and premature infants. *AJNR Am J Neuroradiol,* 27(5), 972-977.

Zhang, L., Thomas, K. M., Davidson, M. C., Casey, B. J., Heier, L. A., & Ulug, A. M.. (2005) MR quantitation of volume and diffusion changes in the developing brain. *AJNR*. 26(1), 45-49.

Zhang, Y. W., Wang, R., Liu, Q., Zhang, H., Liao, F. F., & Xu, H. (2007). Presenilin/gamma-secretase-dependent processing of beta-amyloid precursor protein regulates EGF receptor expression. *Proc Natl Acad Sci U S A*, 104(25), 10613-10618.

In:Cognitive Science Compendium Volume 2
Editors: Miao-Kun Sun

ISBN: 978-1-60456-775-5

Chapter 4

SEMANTICALLY MEDIATED INTEGRATION OF COGNITION IN *HOMO SAPIENS*: EVOLUTION, GRAMMAR, UNCERTAINTY, AND COGNITIVE ACCURACY

Charles E. Bailey
Accurate Clinical Trials, Inc., Kissimmee, Florida
United States of America

ABSTRACT

This article reviews research on human brain, cognition, language, behavior, and evolution to posit the value of operating with a stable reference point based on cognitive accuracy and a rational bias. Drawing on rational emotive, cognitive behavioral and cognitive neuroscience on the one hand and a general brain model of frontal lobe executive function and working memory on the other, along with proposed language mediation of cognitive processes, this review yields potential implications for maximizing brain functioning of *Homo sapiens*. Cognitive thought processes depend on the operations and interactions of specific brain structures and networks, functioning more effectively under conditions of cognitive accuracy (including accurate information, thought process accuracy, and event-level accuracy). However, typical cognitive processes appear to promote the adoption and use of subjective cultural beliefs, mediated by language and grammatical habits mostly learned during early development. In turn, these grammatical habits tend to bias humans toward cognitive inaccuracies. On the other hand, a process that applies informed frontal lobe executive functioning to the mediation of cognition, emotion, and behavior may help to minimize the negative effects of indiscriminately applied cultural belief systems, provide a naturalistic framework for future research and ultimately enhance cognitive accuracy as a reference point for evaluating humans while offering improved relative environmental homeostasis.

Keywords: neuroscience, rational, evolution, grammar, cognition, cultural belief systems.

INTRODUCTION

As an evolving species, *Homo sapiens* tend to use more primitive, inherently inadequate tools to measure the results of thought and behavior. Lacking awareness of how we use words to think and speak, and measuring our success by unquestioned cultural belief systems that we have accepted uncritically, we frequently fall prey to confusion, misunderstanding, and emotional turmoil (Browne & Keeley, 2007, pp. 182-5). Human cognition and behavior developed as a product of evolution, socialization, development, and language mediation, as proposed by Luria, influenced by Vygotsky (Luria, 1981, pp. 1-13) This article contends that application of components of accurate human mental functioning and evaluations, mediated by language, will enhance the probability of more successful, rational outcomes (Bailey, 2006). "Rational" as used here derives from cognitive accuracy and refers to purposeful cognition, evaluation, behavior, learning, and informed deliberation. Rational cognition promotes adaptive, pragmatic, practical, flexible decision-making matched as closely as possible with the present instead of the past. In this view, rationality does not imply finding or knowing a supposed single right answer, but rather recognizes that, lacking omniscience, we do best to prepare for a variety of possible outcomes and adapt readily when things do not happen as we would prefer. "In forming opinion about future events, [rational expectations imply] the use of all available information to assess the probabilities of the possible states of the world. More simply, [rational] expectations [are those] that are as correct as is possible with available information" (Deardorff, 2006). Such flexibility is preferable to "irrationality," used here to refer to behavior that is rigid and reactive, especially cognition oriented towards learned rather than learning behavior, and based on belief rather than evidence, i.e., operating with outdated non-contextual information containing unexamined cognitive inaccuracies that promote incongruent stimulus-bound event-level reactions and primitive objectification.

BRAIN, BONDING, LANGUAGE, CULTURE, AND EVOLUTION

To understand normal human brain functioning and cognition, it may be helpful to distinguish independent variables among organisms, especially between mammals, higher primates, and human primates. What are the variables and how do they affect the species? What does this mean to us as *Homo sapiens*? The largest obvious difference between humans and other primates is the unique prefrontal cortex (PFC) that supports language (Broca, 1877). Subsequent to this distinction, as far as we know, we have more words than any other species (LeDoux, 2002, p. 198). Likewise, words and language make up the largest differences among human populations, forming the basis for relationships and individual cultural belief system values (Luria, 1981, p. 6-7). Humans innately seem to express social attributes for relationships at many levels (Panksepp, 1998, 221-99). Relationships depend on interactions between individuals including cooperation and coordination. These interactions in turn usually depend largely on communication (Cacioppo & Berntson, 2004, p. 978). Faulty, or inaccurate, communication tends to have an adverse effect on relationships at all levels.

Bonding and the Limbic System

We share many of our brain attributes with other primates, particularly our limbic system, which contributes to our lower-level emotional cognition. We can trace the large variation in socialization among species to the differential distribution of oxytocin and arginine-vasopressin (avp) receptors and the pathways that enable bonding, attachment, affiliative behavior, and creation of cohesive groups (Insel, 1997; Bartels & Zeki, 2004; Young & Wang, 2004; Fisher et al, 2002; Cho et al, 1999; Lim et al, 2004). These oxytocin and avp attachment receptors support the bonding cascade that mediates reward pathways in the ventral tegmental area (VTA), demonstrating associations with the nucleus accumbens, bed nucleus of the stria terminalis (BNST), and interstitial nucleus of the posterior limb of the anterior commissure (IPAC) (Heimer et al, 2005, pp. 61-65). These reward pathways add a positive emotional valence, or value, to bonding and affiliative behavior, mediated (at least in part) by a positive shift in VTA dopamine and probably endogenous opioids (Panksepp 1998, p. 263). These pathways appear to overlay lower-level limbic reinforcement and pain pathways identified in mammals and other primates (Tranel, 2000, p. 218). This suggests an evolutionary adaptation of social behavior to preexisting limbic and somatic pain-reward pathways. (See Caldwell & Young, 2006, for a review of oxytocin and vasopressin.)

Unbonding, rejection, social disapproval, or exclusion from social groups (Cacioppo & Berntson, 2004, p. 983) can induce a negative limbic valence, or value. Threats to attachment or approval may produce emotional pain, jealousy, anger, aggression and violence (Insel, 1997). This negative valence appears to correlate with increased amygdala activity. Studies show that the stress induced by detachment or disapproval triggers bereavement-related syndromes, correlated with an increase in stress hormones, corticotrophin releasing factor (CRF), and a decrease in brain-derived neuronal growth factor (BDNF). Other mammals also express increased CRF associated with social defeat and subordination (Ferris, 2006, p. 167-8). This increased CRF in turn correlates with increased amygdala activity and possibly with the downshifting of cognition to implicit limbic and automatic striatal pathways, with decreased PFC and hippocampal volume coupled with diminished executive cognition and attenuated memory. These symptoms echo those of anxiety and depression, sometimes leading to anger and aggression (Panksepp, 1998, p. 205). Shifts in the limbic bonding axis have a dramatic impact on emotional homeostasis, rewarding bonding, affiliation, and approval, while punishing detachment and disapproval. In social animals like human primates, this fundamentally influences cognition, emotion, and behavior (Nair & Young, 2006).

Differentiation, Language, and Socialization

While inter-species social behavior in non-primates and primates may range from very similar to very different, among primates the habit of language distinguishes human primates from our close primate relatives. "Human language offers replication machinery for unlimited cultural evolution", possibly, "representing the biggest invention of the last 600 million years" (Nowak, 2006, p. 249-50). Similar to many social species, humans normally appear to share a genetic predisposition for relationships. For us, however, the innate tendency to bond and affiliate is intimately associated with cultural belief systems, mediated by language

(Panksepp, 1998, p. 245). This underscores the importance and impact of language and communication on relationships, thought, emotion, and behavior (Damasio, 2000, p. 17).

Cultural Belief Systems

The human ability for complex language, formulations, thought, and verbal communication helps define our social interactions (Grafman, 2002, p. 298; Mitchell et al, 2006, p. 63; Risberg, 2006, p. 8-7). Words, grammar, and language support cultural belief systems, which form the major independent variable among normal humans (Luria, 1981, pp. 205-9). Normal human brains exhibit relatively consistent form and function across cultures (LeDoux, 2002, p. 231), but cultural belief systems differ, sometimes dramatically (Whorf, 1956, p. 221). We implicitly learn the structure and rules of what we think and how we think from the culture that we grow up in (Nowak, 2006, p. 263). In other words, communication of cultural belief systems depends on hand-me-down semantics and grammar, which vary from culture to culture (Sapir, 1949, p. 162). Language identifies and defines cultural belief systems and represents a distinguishing group characteristic, directly affecting the thoughts, emotions, and behavior of the group (Adolphs, 2006, p. 269-74).

Cultural belief systems and groups frequently overlap or contain subgroups, but a prerequisite set of beliefs usually determines membership. These systems form within social groups due to our inherited propensity for bonding and affiliations. We find unique belief systems at all scales: individuals, small groups such as families or affiliations, and large groups such as entire societies or states. The rules they embody for the group, regarding thought and behavior, usually pass down from elder members and define the particular belief system. Because the rules and beliefs have direct impact on thought and behavior, they also have a dramatic effect on emotions.

Language, Semantics, grammar, and Cultural Variance

We may assume that a person born and raised in a particular cultural belief system would think, feel, and behave differently than one born elsewhere, not because their brains differ but because they internalize different cultural beliefs (Benjafield, 2007, p. 237; Thompson-Schill et al, 2006, pp. 178-9; Vygotsky & Luria, 1993, pp. 230-31). Even though they share basic human emotions, they will react differently to stimuli based on culturally defined values (Browne & Keeley, 2007, p. 53-69). Words culturally bind our thoughts, our beliefs, and subsequently our behaviors and emotions in a wide range of circumstances (Phelps, 2004, p.1008). Since cultural belief systems rely heavily on semantics—the use and meaning of language—one might also conclude that the use of semantics and grammar figure as the largest independent variables in understanding human cognition and interactions.

These insights highlight the tremendous impact grammar has on emotions, behaviors, and perceptions across cultures as well as between individuals (Boyd & Richerson, 2005, p. 206). Semantics directly affects most aspects of human experience, including cultural belief systems, cognition, emotions, behaviors, evaluations, perceptions, affiliations, pair bonding, bonding mechanisms, social interactions, and even aggressive behavior. It seems reasonable and appropriate to account for this by constructing more *naturalistic* research designs for investigation of human brain function and behavior (Benjafield, 2007, p. 32; Grafman, 2002, p. 293; Delgado, 2007, p. 64; Giesbrecht et al, 2006, p. 104). This will require integration among many fields of science (Roepstorff, 2004, p. 1115; Norris & Cacioppo, 2007, p. 85; Edelman, 1992, p. 252).

Integration of Semantics and Grammar, Biases, and Cognition

Naturalistic approaches seem especially suited to studies assessing complex influences such as the role of higher-order brain functioning on beliefs (O'Doherty et al, 2007, p. 46) and ultimately, on accurate cognition and behavior. Humans have a highly developed frontal lobe system that, along with semantics and grammar, allows for higher-level executive functioning (LeDoux, 2002, p. 197). Even though cognition relies on measured integration of lower-level limbic and automatic cognition (Tranel, 2002, p. 351), the higher-level executive function and higher-order working memory are in a position to have the *last word.* Higher level and higher order as used here refer to executive functioning (DLPFC) with capacity for flexibility, explicit *objective* evaluations and abstractions, as well as considered thought and decision making. We can describe many higher cognitive processes as symbolic processes, including memory, attention, imagery, ideation, concept formation, generalization, abstraction, problem solving, thinking, reasoning, and planning (Logothetis, 2004, p. 849). Lower level and lower order as used here denote more rigid, automatic, stimulus-driven, or subjective emotional limbic appraisals and *subjective* implicit automatic cognition (Benjafield, 2007, p. 44-5, 51, 264-7, 307-9; Frith, et al., 2004, p. 265). This nomenclature allows for parsing of higher level and lower level brain function relative to grammar and cognitive accuracy (see Table 1). In human primates, language supports the option of making decisions based on reason rather than emotions. The last—and possibly the most influential—step in cognitive processing uses information abstracted from our personal history (Stuss et al, 2001, p. 102). This information embodies the relative values of our personal cultural belief system that ultimately biases our choices and their consequences.

From culture to culture, semantic structure, grammar, and the use of language vary. The potential for higher-order abstractions appears enhanced in sophisticated cultures, and this potential seems associated with more highly developed language and advanced education. The processing of complex information gives rise to abstractions mediated by language (Luria, 1981, pp. 26-30). Abstractions range along a cognitive gradient from subjective to objective, depending on the degree of rationality of the grammar and processing. Across cultures, semantic, grammatical, and linguistic structural gradients range from dichotomous to multivariate, concrete to abstract, simple to complex, and few words and concepts to many. Language that supports more complex and abstract thought, at least in theory, seems to impart a cognitive advantage to *Homo sapiens*, "the human that knows they know" (Risberg, 2006, p. 3). By supporting higher-level abstract abilities, language provides the tools with which to think about how we think. Increased use of these higher-level abstract expressions of "thinking about thinking" apparently evolved fairly recently, around the sixteenth century or after (Benjafield, 2007, p. 235). This in turn allows us to discover the more objective, abstract principles of science that can lead to a more accurate understanding of the natural environment, human thought, emotions, and behavior (LeDoux, 2002, p. 176). This scientific understanding then forms the basis for the generation of a cascade of more accurate information for logical problem solving and appropriate objective decision making. Science offers a system for locating knowledge along a subjective-objective spectrum of classification. This distinguishes the subjective knowledge of cultural beliefs, what we *assume* we know, from objective scientific knowledge, what we can *demonstrate* we know.

Table 1. Inaccurate versus Accurate Bias

Lower Level Inaccurate Irrational Bias	Higher Level Accurate Rational Bias
Faulty rigid assumptions; dogmatic beliefs, unsupported by facts, but stated as unquestionable "truths of the Universe" with questioning prohibited, "superstitious" ritualistic thought and behavior that promotes mind-brain dualism, subjective abstractions	Rational flexible assumptions stated as theories; hypotheses and conclusions supported by evidence, scientific testing, and mandatory questioning, "scientific" adaptable thought and behavior, mind = brain = mind/brain, objective abstractions
Rigid, maladaptive, with lower-level subjective bias, vertical subordinate communication	Flexible, adaptive, with higher-level objective bias, collateral communication
Absolute, static bias: certain, *"determinate,"* guaranteed	Variable, dynamic bias: uncertain, *"probability,"* not guaranteed
Cognition using dichotomous grammar limits freedom of executive function	Cognition using multivariate grammar expands freedom of executive function
Veridical bias: true and false, either-or, absolute, concrete, black and white; constrictive and restrictive, *not contextual*	Associative bias: abstract, gray, gradated; expansive and extensive, *contextual*
Predetermined certainty, all knowing, resulting in decreased frontal lobe requirements: *"afrontal"*	Relative uncertainty, inquisitive, resulting in increased frontal lobe requirements: *"frontal"*
Parental, demanding; adversarial	Adult, requesting; cooperative
Semantic inaccuracy: vague, poorly defined, with overgeneralizations: always, never, every, all, none, etc.	Semantic accuracy: specific, best definition and word use: frequently, infrequently, many, some, few, etc.
Rigid, implies no other choices: I should, I must, I have to, I need to, and I have got to. *"I am obligated"*	Flexible, implies choices; preferential: I prefer, I would rather, I would like to, I choose to. *"It is a choice"*
Tends to ignore inaccuracies of information, of thought process, and of event-level orientation; retroactive, "reactive"	Tends to promote accuracies of information, of thought process, and of event-level orientation; forward-thinking, proactive, "considerate"
Inaccuracies and faulty assumptions promote faulty and inaccurate cause-and-effect conclusions	Accuracies and rational assumptions promote more plausible and more accurate cause-and-effect conclusions
General unawareness of irrational cognitive process "Cultural belief system anosognosia"	General awareness of rational cognitive process and "Cultural belief system awareness"

When it comes to understanding humans and the human brain, it also seems important to consider the significance that semantics, grammar, and the resulting *beliefs* and biases have for human evaluations and interactions (Jones, 1998, p.14-23). *Brain Res*earch lags in consideration of the subjective cultural bias of semantics and language, and it rarely accounts for the relative semantic and grammatical accuracy of cultural belief systems and their relationship to thought, emotion, and behavior. Perhaps the relationship often goes unnoticed because we implicitly take our own belief systems so much for granted. The automatic semantics and grammar of our cultural belief systems lie at the center of our oldest and most strongly registered memory traces. They mostly operate implicitly—their day-to-day causal effects generally invisible to us—and we often take them at face value, without notice or challenge. Indeed, we simply and spontaneously tend to accept the grammar and beliefs of our culture as *factual* (Benjafield, 2007, p. 400). Unexamined inaccurate cultural beliefs, however, directly contribute to the mechanical use of inaccurate information along with rigid, dichotomous inaccurate cognition (Hooker & Knight, 2006, p. 317).

Evolution of Grammar

Human grammar contains many hidden, habitual attributes that directly affect our cognition, but we use them on a daily basis without inspection or consideration of reliability or accuracy. These hidden features include primitive, unscientific, subjective, and over-generalized object classifications, categorizations, groupings, and labeling. A relative comparison of language "grammar" equivalents between mammals, other primates, and human primates shows many similarities in basic object-action cognitive processing, demonstrating the high level of evolutionary conservation in mammalian brains. Brain areas for object-action cognition tend to function similarly in mammals and other primates, providing representations of stimulus pattern characteristics including object identification, object spatial location, and limbic mediated object risk-reward contingency value including evaluating intention of animate objects. However, human primates possess large visual and auditory association areas that represent recent evolutionary development for specialized vision and language processing networks. Theories of innate universal grammar postulated in linguistics may have derived from observed pattern recognition functions in the object-action-intention predisposition in mammals, primates, and human primates to the relative evolutionary conservation of brain function (Hauser, Chomsky & Fitch, 2002). However, a usage-based approach used here seems to offer a simpler model with more ecological validity for cross-cultural grammatical appraisals (Tomasello, 2004, pp. 642-5); Luria, 1981, p. 6), especially relative to evaluating the functional differences between multivariate and dichotomous grammar.

Language incorporates the skills of speech and grammar to provide a basic substrate for higher-level object and action description and identification, cognitive and emotional processing, and communication. Grammar also incorporates other sensory components, adding further descriptive value to stimulus object-action perception. Other primates have rudimentary language components but lack the sophistication of human syntax and grammar leaving their communication literal and mostly inflexible by comparison (Mesulam, 2002, p. 19). Mammals including other primates rely on sensory information to regulate environmental homeostasis, while limbic components provide valuable subjective risk reward contingency, information for actions. Homeostasis relies on feedback to maximize adaptation over time. This feedback may largely depend on semantics and grammar in humans. We have the benefit of words and grammar to assist in evaluating object salience and intention, contingency value, as well as action choice and planning. Contingency implies prediction. Prediction of causality requires awareness of the contingency between the intention and the action and between the action and its consequences (Portas et al., 2004, p.288). The accuracy of this awareness would seem pertinent for obtaining the most objective evaluations. Rigid limbic-driven imperatives dominate most other animals (Mesulam, 2002, p. 19). It seems that in humans, even though we have the added flexibility of our more sophisticated brain, we often simply augment or replace the limbic imperatives with rigid subjective semantic imperatives, especially regarding our subjective inferences and perception of the intentions of others. Humans have the option of lower level subjective limbic evaluations and higher-level objective evaluations not afforded to other mammals.

Vision and visual processing occupies a large part of the human brain. The addition of grammar creates value by providing a mechanism for parsing the categorization of environmental stimuli into discrete categories of objects, groups of objects, intentions, and actions (Nowak, 2006, p. 251-2. Vision plus grammar maximizes environmental object detection, subsequent labeling, and identification. Auditory detection and processing provides value, especially regarding the listening component of communication. The addition of grammar to motor function, the evolution of speech, and later writing, provided important components of communication. Symbol use and speech probably evolved around 50,000 to 100,000 years ago, while writing most likely evolved within the past 4,000 years (Striedter, 2005, p. 312-13). It seems to follow that early in language evolution, oral communication provided the major component of cultural transmission of information. Since this predates writing, story telling by narratives, influenced by implicit and explicit recollections and emotions, most likely provided the medium for informing each new cultural generation (Siegel, 1999, p. 333). In many cases, cultural belief systems, anecdotes, magical interpretations, myths, and story telling continue to trump scientific facts and statistical probability. This might also explain the prevalent human propensity to rely on primitive dichotomous grammar constructs and imprecise pattern recognition principles supported by superstitions, primitive rituals, over-generalizations, faulty subjective information, and faulty cause-effect conclusions (Vygotsky & Luria, 1993, pp. 138-9).

Cognition, an important piece of the human brain function puzzle, in large part relies on grammar. Evolutionary theory suggests that we benefit most as a species by achieving and maintaining consistent homeostasis with our dynamic environment. Furthermore, we can assume that we achieve the most effective object categorization and recognition, which guides our subsequent actions, by integrating and processing information pertinent to our event-level relationship with our environment. From this, we can conjecture that human long-term overall survival and success might depend on our ability to obtain the most accurate identification of objects and stimuli and the most accurate internal information about object relationships, and to process this information with the most accurate processing rules we can develop. To the extent that we can do this, we learn to make the most accurate assessments and devise the most reasonable methods for integrating variables involved in problem solving, for developing reasonable strategies, and for choosing the best actions to achieve a desired goal.

Evolution of Subjective Cultural Classifications

We might assume that language and grammar have evolved over evolutionary history as an adaptation for maximizing homeostasis of *Homo sapiens* with its environment, which includes other *Homo sapiens*. The phylogeny and ontogeny of words and grammars across cultures and history appear to offer some clues to the evolution of this adaptation. Words as linguistic symbols adapted by humans have evolved along side the social culture, as the defining language of the culture and the concomitant cultural belief system. The conventional usage learned by children includes the words and formulation forebears found useful in the past. Grammars evolve over many generations as our ancestors developed words, meanings, and grammar structure in response to changing environmental factors over time. This evolution developed from concrete utterances and symbols operating through cultural historical processes, rather than biological ones. Grammars passed along to subsequent generations, succeed as a product of familiarity, exposure, and the potential practical value they afford each new generation (Tomasello, 2005, p.13).

We might expect that the higher-level skills of more accurately identifying, manipulating, and interfacing with objects and other *Homo sapiens* in our environment would provide humans with survival benefit. Language offers a socialization advantage unique to humans, enhancing our ability to communicate, share information, plan actions, and understand the intentions of others (Risberg, 2006, p. 10). Nearly all language provides nouns for referentially labeling objects, and verbs for predicatively labeling actions or behaviors. It seems to follow that the more accurately we describe objects and predict their likely actions, the greater our ability for conceptualization of abstract representations. This improves our decision–making abilities, which in turn confers a greater adaptive advantage. Complex languages that enable much more objective higher-level abstract formulations would seem to offer an even greater adaptive value. However, as grammars become more complex and abstract, they can become less precise, with a potential for increased variance and errors. The importance of error monitoring, detection, and correction then increases. This suggests that assessments of abstractions based on accuracy might offer more precision and fewer errors, i.e., more objective, scientific abstractions versus more subjective abstractions.

Language evolved from primitive nonliterate cultures dominated by uninformed abstractions and subjective ways of knowing, and language still appears to convey some of that imprecision in cultures that continue to cling to subjective beliefs as facts. Science represents the other end of the knowledge spectrum, as demonstrated by the acceleration of more objective knowledge, i.e. scientific facts and technological advancements. The continuing dynamic seems to highlight the opportunity for rational intervention and the establishment of a more accurate reference point for measuring the objective descriptive precision of grammar, semantics, and abstract conceptualizations. Many of the abstractions humans use on a day-to-day basis fall into the category of *subjective abstractions* with inherent inaccuracies and irrational biases. *Objective abstractions*, on the other hand, use scientific principles and critical thinking skills to maximize accuracy, resulting in a more rational bias. This distinction, between cultural subjective categories and processing versus scientific objective categories and processing, offers a potentially beneficial reference point for human evaluations.

Ontogeny and Phylogeny of Grammar

How could humans evolve such a powerful dynamic evaluative capability as the human modern brain and continue to operate largely on primitive subjective evaluations? Infants begin their interactions with the world with generalized expressions, gestures, and utterances including cries, grunts, and coos. These and the concomitant facial expressions represent the primary form of communication with caregivers. Over time, infants begin to add single words and abbreviated short word phrases, as they master the naming of objects and actions, leading eventually to predication.

Verb development usually follows noun development. Labeling objects with names appears to help the child master the environment, and children seem biased toward using any new word as an object name (Bloom, 2002, pp. 92-9). After the child learns to use verbs and later to construct whole phrases, this functional vocabulary provides the foundation for the exponential growth for grammar development. Even in nonliterate cultures, the selection of objects to label reflects the linguistic ideology of the culture. Children apparently learn most of their early vocabulary simply by listening to the conversations around them. This applies particularly to preliterate children and to older children and adults in nonliterate societies (Bloom, 2002, pp. 118-9, 192). Much of this early vocabulary represents the building blocks of the background cultural knowledge and categories that structure and determine word meanings for abstractions related to space, time, causality, objects, intention, and possession (Tomasello, 2006, p. 54). This cultural knowledge also appears to bias the cultural frame of reference used in various other representations (Bloom, 2002, p. 247).

Children deduce early in language development that when adults refer to objects, they do so in terms of whole objects (Bloom, 2002, p. 92), i.e. they tend to over-generalize. A major part of human linguistic competence involves mastering by rote many routine formulas, fixed and semi-fixed expressions, idioms, and frozen collocations that objectively have somewhat unpredictable or inconsistent meanings (Tomasello, 2006, pp. 101-2). In many languages, including English, these expressions represent templates that, when used in conjunction with a coupler, can generate an almost infinite number of culturally *subjective* abstractions. A common grammar component across many languages is the copula, or coupler. In English, the verb "to be" represents the primary copula, and its use appears early in development (Tomasello, 2006, p. 255). Unfortunately, despite its ubiquity, "to be" is generally insensitive to pattern specificity, time specificity, and context specificity, which makes it very useful for subjectively over-generalizing labels but at the cost of reduced precision and distorted perceptions.

The utility value of the predicate form of "to be" encourages subjective generalization of behaviors (actions) to objects. This allows indiscriminant objectification of dynamic processes by semantically converting any subjective condition into a seemingly objective noun (i.e., we can say, "He *is* a *failure*," instead of saying, "He *failed* at this particular effort" or "She *is bad*," rather than "She acted *badly* last night"). The verb "to be" represents great utility as a very handy general label-making tool. In light of this generalization utility, "to be" might represent the basis for primitive verb evolution. Cultural labels often represent subjective, discriminatory, rigidly held prejudices, biases, and habitually defended beliefs, supported in lieu of objective scientifically derived classifications or alternatives. Objectification, or subjective labeling of humans or groups of humans as objects, presents a biased, usually derogatory "image" of those humans. Human adults and infants reflect this bias (Bloom, 2002, pp. 93-4). Other evidence suggests that the mere presence of labels may encourage people to exaggerate differences between groups (Bloom, 2002, p. 254).

Cultural belief systems also rely on prescriptive, imperative, rigid, authoritarian, and sometimes intimidating auxiliary verbs, or "helping" verbs (should, must, have to, need to, ought to, got to, etc.) that do not convey contextual specificity. Descriptive verbs describe the reality of the world, while prescriptive verbs describe how the world *should be*, usually based on unspoken assumptions. Unfortunately, prescriptive verbs usually express a subordinate frame of reference bias held by the *should-er*, but not necessarily known or accepted by the *should-ee*. "Need" is often overused subjectively. Humans need food, water, shelter, and air to survive. We may want or desire other things but they hardly meet objective criteria as needs. Many of these prescriptive and imperative verbs represent arbitrary cultural artifacts that deftly sustain the cultural belief system and exclude rational choice or consideration of reasonable choice-outcome paradigms. Each decision has many choices and variable consequences. Prescriptive and imperative statements demand a predetermined *choice*, with insensitivity to awareness or consideration of the *context* or the nature of the decisions and the probable outcomes involved (Capaldi, 1987, p. 17). Prescriptivism supports dominance grammar and consequently supports and sustains cultural belief systems.

Factitive and causative verbs combine a direct object and a phrase to apply a certain characteristic or a change in status to an objective complement. One such verb, "to make," works hand in hand with prescriptive statements. "You *make* me sad, you *made* your father angry, you *made* me act that way." These subjective verbs enable an individual or group to transfer to others the responsibility for their own thoughts, emotions, behaviors and choice-consequence outcomes. Factitive and causative verbs probably made sense historically in the primitive evolution of grammar, language, culture, and learning. Perhaps due to their social utility, they survive even in well-educated cultures by providing a vehicle for maintaining cultural belief systems and superstitions, transferring responsibility, and blaming of others.

Humans also often apply faulty knowledge and confuse coincidence or correlation with cause and effect (Skinner, 1953, p.84-5). We learn these habits as children, by listening to the conversations of adults. Unfortunately, "children aren't like scientist who have theories; they are like scientist before they have theories, trying to make sense of some domain they know little about" (Bloom, 2002, pp. 168-9). Children physically grow into adult humans but generally retain their subjective cultural belief system biases. Science requires inferences based on scientific inquiry with objective observation, and correlations with statistical probability. "Science attempts to find out how things really are, not just how they appear to be." (Bloom, 2002, p. 169) Subjective cultural belief systems often result in this true or false, dichotomous grammatical underpinning to human cognition.

Dichotomous grammar, consisting of veridical, either-or, black or white structures, continues this pattern of culturally learned conventions. Culturally derived dichotomous cognition automatically preempts objective, multivariate cognitive strategies for evaluation, even in adult humans. The verb *to be*, imperative prescriptivism, and either-or cognition represent complementary candidates for the primary source of primitive dichotomous grammar evolution. Even though these habits become automatic and implicit in a sense, they can become apparent with ongoing effort and inspection. However, we have little chance of noticing them, finding them, or correcting them if we do not even think to look in the first place (Jones, 1998, p.33).

Evolution and Dominance Hierarchies

An evolutionary model seems to offer some plausible explanations for the relationships between culture and language. If we start with the assumptions that dominance hierarchies have historically supported the evolution of adaptability in human social systems, and that survival depends on group functioning, then it seems to follow that hierarchies controlled through reinforcement and punishment by dominant individuals or groups of individuals have historically conferred an adaptation for survival and reproduction. This applies to many other species as well as to humans. Indeed, in modern human interactions and language, we may still observe the slow evolution of these fundamental dominance features. If language and grammar have evolutionary characteristics, we might benefit from considering how these adaptations evolved.

A language, in some ways, resembles a species, as it defines social groups. Within most species of languages, there is variety, and most large groups exhibit significant differences in dialects and colloquialisms. Greater descriptive variety and complexity in language and grammar can enhance higher-level abstractions and flexibility; in contrast, constricted or subjective dichotomous homogeneity shifts the speaker towards concreteness and rigidity. If descriptive complexity enhances the potential of language features, including higher-level abstraction, and offers an adaptive value, we would expect to see languages with these features proliferating over time and becoming more dominant; indeed, this seems to be the case. Descriptive variety appears to endow language with this adaptive evolutionary trait of diversity, in much the same way that variety and diversity supports the natural selection processes described by Darwin in *The Origin of the Species*.

This spectrum of linguistic and grammatical variety, ranging from concreteness to abstraction, represents various gradients used for cognition: from static thinking to plastic, rigidity to flexibility, dichotomous to multivariate, resistance to diversity and change to enhancement of diversity and change. We observe a similar evolution in organisms from single-cell simple systems to multi-celled complex systems, with each level of complexity usually enabling enhanced flexibility and adaptive value. Cultures tend to parallel this progression from simple to complex, brought about and supported in large part by complex language and grammar. If we apply evolution theory to grammar and social development, we may assume that the *beliefs* that helped our ancestors and parents reach the age of reproduction and produce offspring may, in a sense, pass on to their children as a inherited adaptive value, however rigid or inaccurate. When contemplating how languages and cultural belief systems could possibly demonstrate such rigid resilience over time, it helps to remember the span of human evolution represents a very brief time evolutionarily. In our earliest stages as humans, we most likely benefited from a strong sense of confidence in the historical success and survival of the group. In this brief window of social evolution, in some ways we see relatively small transmutations from rigidity to flexibility, while our brains have progressed biologically to the point where they can support abundant flexibility. The firmly ingrained language of social rules and beliefs, supported by rigid grammar structure, seems to impose impediments to change.

Indeed, dominance hierarchies in humans—characteristic of our genetic inheritance and conditioned by our environment—have a large influence on this stodgy rigidity. Our brains contain hard-wired circuits for aggression, territoriality and competition for resources, dominance status, parental-familial-affiliation, defensiveness, and irritability (Wingfield et al,

2006, p. 180). These circuits represent a significant inertia and resistance to the evolution of more objective and rational language use that can replace our typical irrational, dominance-related expressions of aggression. From individuals to whole cultures, we see dominance and rigidity perpetuated by the grammatical language habits acquired in early development. Our learning mechanisms (including *modeling*, instrumental learning, and associative conditioning) largely depend on implicitly embedded language and grammar, which makes it difficult for us to discover and overcome the inertia of our belief systems. This learning, supported by semantics, forms the basic template of *what we believe* and *how we think* for most of our lives. Our language and grammatical processes provide the vehicle for culture, social structure, and adaptation that, theoretically and potentially, enable us to operate on a gradient from a subjective irrational bias to a more objective rational bias. Culturally, we tend toward the former, while cognitive accuracy offers a way to move closer to the latter.

Cultural Social Inheritance

Through the mechanisms of genetics and learned belief systems, our thought and behavioral patterns reflect our personal, social, and cultural histories. Genetically, we inherit a basic blueprint for how our brains operate. Culturally, we acquire a blueprint for what we think, how we think, and for expected behaviors. Inevitably, as humans we inherit cultural beliefs representing a long line of learned and rigidly held subjective inaccuracies that bias our perception. Becoming aware of the potential inaccuracy of what we know, *or think we know*, allows us to make corrections and to think more critically. We measure cognitive accuracy by the relative distance or gradient between the unexamined, inconsistent, and irrational yardsticks we have inherited and the validated, external, more reliable, rational reference points we have identified through science. The shorter this distance, the more objectively, rationally, and accurately we think.

Most people contend they think accurately, rationally, and logically. However, they generally base their contentions on their own usually unexamined, inaccurate, and irrational frame of reference. For the most part, our individually inherited grammar and cultural belief systems subjectively bias our thoughts and perceptions, even of ourselves. This bias becomes apparent when compared with a rational reference point or standard (see Table 1). Awareness of this irrational bias opens the door to the adoption of standards with more objective accuracy and reliability, resulting in the potential for a more objective rational bias.

Human Brain Model

Neuroscientists sometimes describe normal human brain functioning in terms of a computer model (Avrutin, 2006, p. 49; Benjafield, 2007, p. 14-18; Braver & Ruge, 2006, p. 338-9; Neisser, 1967; Panksepp, 1998, p. 20; Shiffrin & Atkinson, 1969; Toates, 2007, p. 17; Baum, 2004, p. 7). Our brain (*hardware*) comes initially from our inherited genetic blueprint; environmental learning makes the major contribution to our stored memory information (data), and grammatical process information (*software*) (Schmalhofer & Perfetti, 2007, pp. 180-81). The brain stores memories as information in various storage areas, similar to a computer's hard drive. A portion of this storage contains grammatical templates for *processing* information—the rules of how we think, similar to computer software that determines how information is processed (Fuster, 2003, p. 55). Like computers, humans may acquire or develop faulty or pathological hardware, software, or data. Both computers and humans get the most accurate results with the most appropriate hardware, most up-to-date adaptive software, and as accurate and timely data as possible. This enables the most reliable choice-outcome conclusions at a desired point in time. Neither computers nor humans can produce satisfactory results with inaccurate, out-of-date data or faulty, inflexible software. We tend to build error detection into our computers, and to upgrade and install new versions of software as new information and technologies become available — it seems logical to look for ways to do the same with our own human brain computers. If instead we approach the brain as a black box with only input and output, we would most certainly have very much interest in the unknown processing going on inside the black box. Fortunately, we do have a growing knowledge of this processing in the human brain. The quality and precision of the output seems to correlate with the quality of our information and quality of its processing. With direct evidence of the adverse affects of irrational grammar usage on cognition, emotions, and behavior, we can abandon the ancient perception of the brain as a mysterious black box, and scientifically apply what we have learned about it to improve the way we make use of it.

Frontal Lobe Integration: Executive Functioning and Working Memory

The higher-level executive working memory (Baddeley, 2002, p. 246) of the brain's frontal lobes (specifically, the integrated functioning of the dorsolateral prefrontal cortex (DLPFC), along with frontopolar cortex (FPC), Broca's area, temporal, temporoparietal and association areas, etc.) compares to a computer's random access memory (Mesulam, 2002, p. 26). The information the brain uses to make decisions with compares to the data stored in the computer. The DLPFC and FPC appear to play an important role in the integration of internal and external appraisals to adapt to changing conditions (Gazzaley & D'Esposito, 2007, p. 188; Risberg, 2006, p. 6; Wagner et al. 2004, p. 714) and with explicit empathy with others (Decety, 2007, p. 258-60; Ferstl, 2007, pp. 87-90). This dynamic integration of information enables critical error monitoring, error detection, and error correction. The left DLPFC plays a large role regulating language function and sequencing while the right DLPFC has more involvement with the processing of interpretations, inferences, and concepts, awareness of novelty, situational and emotional appraisals, as well as autobiographical memory (Schmalhofer & Perfetti, 2007, p. 184; Long et al, 2007, p.330; Tapiero & Fillon, 2007, p. 365; Siegel, 1999, p. 331). The OFC assists with a parallel role in mediation of emotion and object-affect associations (Mega et al, 2001, p. 23).

The anterior cingulate cortex (ACC) appears to play a pivotal role in motivation-related error detection (Niv, 2007, p. 369), novelty/complexity detection and performance monitoring (Braver & Ruge, 2006, p. 322-24) by interfacing with the DLPFC and limbic, temporal lobe, parahippocampal gyrus, and automatic subcortical association pathways (Kaufer, 2007, p. 49-52). When automatic pathways offer incongruent motivational resolutions to homeostasis, the DLPFC generates congruency. Over time, "congruence strategy" efficiency will benefit from up-to-date information, continual monitoring and improvement of error detection strategies and increased error resolution precision. Error correction may be enhanced deliberately and rationally, by updating the accuracy of perception i.e. information, process, and event-level accuracy, or reflexively and irrationally, by explaining away discrepancies and preserving subjective beliefs and perceptions despite their inherent inaccuracies. Without frequent evaluation of the relationship between our behavior and the environment relative to reality, we can easily fall prey to stimulus bound reactive behaviors (Luria, 1966/1980). As a part of the broader network involved in adaptive decision-making (Lee & Seo, 2007, p. 108), the DLPFC can conduct more objective, considered, and deliberate routine monitoring (Goldman-Rakic, 1995) of appropriate cognition, emotion, and behavior (Braver & Ruge, 2006, p. 321), perhaps enhanced by operating with a perspective biased toward accuracy and objectivity. This objective accuracy bias potentially offers improved error detection and correction with enhanced risk prediction (Preuschoff & Bossaerts, 2007, pp. 142-45).

Even though the orbitofrontal cortex (OFC), limbic system, hippocampus, striatum, etc., play a critical and often parallel role in working memory and decisions (Wagner et al, 2004, p. 709), "there is a hierarchy" (D'Esposito & Postle, 2002, p. 177; Curtis & D'Esposito, 2006, p. 295, fig. 9.8). "Executive functioning requires the integration of prefrontal and subcortical activity", (Mega & Cummings, 2005, p. 18). Semantics, grammar, and language seem to contribute, directly or indirectly, to accurate cognition at each level. A series of parallel networks, including the subcortical cognitive pathways, provide information for executive decisions and explicit and implicit automatic cognition, allowing for adaptive interactions with environment (Chow & Cummings, 2007; Salloway & Blitz, 2002; Lichter & Cummings, 2001, Middleton & Strick, 2001). This hierarchy allows for top-down use and regulation of subcortical systems to enhance monitoring, and correction of contingent input-output incongruence (Stuss, et al, 2001, p. 101). An anatomical and functional viewpoint of these networks seems to offer some relevance for understanding this hierarchy.

The frontal lobes comprise two basic anatomical and functional systems (Pandya & Barnes 1987), the dorsal system consisting of dorsolateral and medial portions of the frontal lobes, and a ventral system consisting of the orbital portions of the frontal lobes. The dorsal system has interconnections with the posterior parietal lobes and cingulate gyrus, and deals with sequential processing of sensory, spatial, and motivational appraisals of external environmental objects and stimuli (the *where* of the event). The ventral system has interconnections with limbic networks involved in regulation of internal homeostasis and emotional conjectures about perceived external objects (the *what* and *why*, or intention), as well as appraisals of stimuli salience, valence, and motivational relevance (Ogar & Gorno-Tempini, 2007, p. 59). The two basic divisions along with their two primary subdivisions are:

- DLPFC, providing executive functions of deliberate, explicit executive cognition/explicit working memory involved in processing higher-level cognition. The superior medial prefrontal cortex (a.k.a. ACC) functions as an automatic explicit/implicit integrative center for cognitive-behavioral (attention-motivation and possibly error monitoring) and emotional-autonomic-motor neural networks (Chow & Cummings, 2007, p. 25; Kaufer, 2007, p. 49).
- OFC, providing limbic regulation and ACC interaction involved in emotional integration of automatic, explicit/implicit, emotional working memory (medial OFC), including ventral striatal, medial temporal lobe (MTL), amygdala, hippocampus, and hypothalamus. In addition, the lateral OFC is involved in

social working memory processing for emotional components of social motivation and cognition (Chow & Cummings, 2007, p. 30-1; Salloway & Blitz, 2002, p.10; Mesulam, 1985).

Each of these systems may contain subsystems along with accessory parallel frontal-subcortical circuits (FSC), consisting of cortico-striato-pallido-thalamic-cortical loops. There have been at least five to seven proposed general loops including skeletomotor, oculomotor, dorsolateral prefrontal, anterior cingulate, lateral orbitofrontal, medial orbitofrontal, and possibly inferotemporal/posterior parietal (Alexander et al., 1986; Middleton & Strick, 2001). Each of these loops appears to be comprised of multiple parallel segregated circuits. These pathways offer multiple levels of parallel processing of information for cognition, emotions, and behavior, including extrapyramidal, motor, and speech systems probably including rule-based cognitive processing (Poldrack & Willingham, 2006, p. 130-1). The basal ganglia have direct input from the large association cortex in the parietal, temporal, medial, and frontal areas, as well as the hippocampal formation and the amygdala. There is also large outflow directed towards the frontal cortex via synaptic links in the thalamus, outflow to temporal and parietal lobes linked to executive functioning (Graybiel & Saka, 2004, p. 495-7) and open pathways to inferotemporal and posterior parietal areas.

In addition, other pathways provide integrated reflexive, or innate, lower-level automatic, stimulus-response pathways, responding to potentially threatening stimuli, i.e. loud noise, fast movement, gestures, as well as to threatening facial expressions and involvement in reflexive empathy. The frontal lobes evolved to manage the pyramidal pathway and to implement cognitive decisions in adaptation to a dynamic environment by incorporating a wealth of sensory input, using higher cognitive function to prioritize that input, and choosing the most effective response, by ongoing evaluation of shifting priorities, initiating action, and monitoring its execution (Chow & Cummings, 2007, p. 38).

The separate but parallel dorsal and ventral anatomical systems appear to represent two functionally distinct hierarchal networks relative to objective higher level DLPFC and subjective lower level OFC cognition respectively. "'Executive cognitive functions' are defined as (and the term is limited to) high-level cognitive functions, believed to be mediated primarily by the LPFC (lateral prefrontal cortex), that are involved in the control and direction (e.g., planning, monitoring, energizing, switching, inhibiting) of lower level, more automatic functions" (Stuss, 2007, p. 293). Executive functions include the ability to solve a complex problem and organize a volitional behavioral response in a temporally informed manner. They also encompass the learning of new information, the systematic searching of memory, activation of remote memories, appropriate prioritization of external stimuli, attention, generation of motor programs, metacognition, and probably the use of verbal skills to guide behavior (Chow & Cummings, 2007, p. 29; Cummings & Miller, 2007, p. 15). Language, semantics, and grammar usage associated with these distributed neural networks may explain the integration of orbitofrontal/emotional and dorsolateral/cognitive processes and resolve the long-standing difficulties in the reconciliation of internal integration and processing of cognition and emotions, especially for higher order processes such as decision making and social perception (Gazzaley & D'Esposito, 2007, p. 190-1).

The DLPFC modulates cognition and human environmental interaction from the top down via integration of externally driven information with internally represented information. The relative effectiveness of cognition seems to depend on the quality of external information received or perceived, and on the perceptual quality of information stored in long-term memory, including "beliefs" related directly to information, information processing, emotions, responsibility, and intentional perceptions of others (Gazzaley & D'Esposito, 2007, p. 192-7). Language may directly mediate the quality of cognitive integration as measured by objective accuracy. Executive functioning depends on the integrity of instrumental functions such as language (Cummings & Miller, 2007, p. 16). Semantics and grammar appear to act as the common denominator between higher-level executive functional networks and lower-level limbic system functional networks, supporting appraisals for the identification and perception of object/stimulus salience, emotional valence, actions, and reactions.

Human evolution seems to have "thoughtfully" overlaid semantic representations onto our cognition. Words and concepts bind to sensory information, creating a kind of a sixth sense—specifically, a semantic sense (LeDoux, 2002, p. 203). This allows language and grammar to provide a substrate for human cognition using words to assign values to stimuli, emotions, behaviors, perceptions, belief systems, etc. Culture provides the foundation for the development of values by grounding specific cultural beliefs in grammatical semantic usage, coupling words and beliefs with historically learned reward contingencies during development. Words have become a sense for detecting and coding predictive relationships among stimuli, objects, responses, and rewards. As such, words assign the relative emotional and motivational significance for the value of rewards and the anticipation of rewards (Phelps & LaBar, 2006, p. 432). These evaluations are automatically mediated by the OFC (Petrides & Pandya, 2002, p.45-46) and depending on task demands, will often default to lower level automatic grammar habits managed by the striatal complex (Poldrack & Willingham, 2006, p. 134-40). Cultural beliefs thus underpin the fabric of human social interaction, by relying on learned, automatic implicit grammars that allow speech to flow effortlessly. Mostly learned by implicit inductive inference in childhood (Nowak, 2006, p. 263), habitual grammar forms a silent, "ritualized" soundtrack for our adult cognition.

Words and grammar give us a unique descriptive ability to encode environmental cues with the potential to maximize efficient usage of neocortical associations. Depending on task demands and stimuli, association mechanisms generally run implicitly and almost continuously in the background, evaluating object features and place associations, risk-reward associations, and past related choice-outcome associations. As complexity, novelty and unfamiliarity increase (Anderson et al, 2002, p. 505), and when higher-level abstract demands increase, the brain typically achieves more preferred outcomes by using the executive functions of the DLPFC and FPC (Curtis & D'Esposito, 2006, p. 293-95; Braver & Ruge, 2006, p. 330). However, in routine situations, lower level limbic and habitual cognition often suffices.

The lower level OFC provides limbic regulation and valuable utility calculations (Glimcher et al, 2005) while monitoring somatosensory information, emotional valence, relative risk-reward contingencies, and social salience. It also contributes to inhibition of inappropriate responses (Rolls, 2002, p. 370). Except for the sophisticated language system, the human OFC somewhat resembles that of other primates. However, in humans, the OFC also automatically provides ongoing bottom-up regulation of the limbic system using the default semantics, language, and grammar from our cultural belief system, thus reinforcing the subjective social, neuroeconomic, and relative psychological values that define our cultural bias (Kahneman et al, 1982; Padoa-Schioppa & Assad, 2006). When subjective cultural belief system grammars are edited to operate more accurately (i.e. yielding a more scientific belief system), the bias shifts towards the more objective end of the value probability continuum, operating closer to the rational-cognitive end of the evaluation scale rather than the emotional-motivational end (Schultz & Tremblay, 2006, p. 195).

Information Content and Information Processing

The top-down executive functioning of the DLPFC likewise relies heavily on semantics, language, and grammar to perform critical integration of choice-outcome determinations and inhibition of inappropriate responses. The lateral prefrontal cortex enables greater complex abilities and flexibility in human and primate behavior (Striedter, 2005, p. 309). The newer higher level DLPFC adds the potential for more accurate rational thought onto the older lower level limbic brain by enabling humans to reason (Panksepp, 1998, p. 20). Efficient prefrontal cognition relies on the quality of information received from other cerebral regions (Anderson et al, 2002, p. 506). While rigid cognitive inaccuracies can impede the discovery and executive execution of the most logical and reasonable decisions, shifting toward an accurate and rational adaptive bias enhances objective cognition, enabling more reasonable choices and outcomes, emotions and behaviors. In terms of the somewhat simplistic computer analogy, the DLPFC executive function, probably aided by information integration involving the FPC, has the capacity for objective appraisal, reappraisal and flexible editing, while the OFC is limited to subjective emotional-based read-write operations, or subjective contingency appraisal and reappraisal. Evidence indicates that task relevant semantic knowledge develops in a relatively automatic (bottom-up) fashion or in a more controlled (top-down) manner (Wagner, et al., 2004, p. 715) probably flexibly regulated by the DLPFC. The DLPFC sits at the top of the PFC as the captain of the ship, demonstrating a higher level of competency when provided with the resources of the most accurate and the most up-to-date information.

Even though the limbic working memory can adjust responses, it does so primarily based on reward contingencies, reinforcement, and emotional valence spectra (Frith, et al, 2004, p. 265). The OFC adds some subjective flexibility, but our ability to objectively change and integrate the accuracy of *what we know* and *how we think* relative to the environment depends most heavily on the reappraisal and edit functions of the executive DLPFC working memory. The DLPFC compares internal stored memories and habits with external, objective, up-to-date information, with the rational capacity to temper our cognition, emotional responses, and automatic behavioral and grammatical responses to stimuli (Mesulam, 2000, p. 93). The DLPFC has the potential to edit our culturally inherited grammar and provide improved cognitive accuracy for error detection and correction. In addition, the FPC (BA 10) appears to have strong connections with temporal areas that support auditory working memory (Barbas, 2006, p.51; Davachi, et al., 2004, p. 670).It appears that, by holding information online during secondary task manipulation in working memory, the FPC supports DLPFC edit functions (Petrides, 2005).

The DLPFC appears to have the ability to directly select or reject individual strategies and may possibly edit grammar through direct mechanisms. However, the edit function most likely relies on explicit DLPFC-initiated reentrant "working memory" rehearsal loops for corrections (Curtis & D'Esposito, 2006, p. 283). These loops develop by practice according to Hebbian rules (Hebb, 1949), and the information at some point becomes routine and then automatically provided "preferentially" by cortical association areas (Braver & Ruge, 2006, p.

328). Corrections might also occur over time due to environmental contingencies, modeling, and instrumental and conditioned learning. Cognitive therapies take advantage of this edit function by using explicit rehearsal to teach patients more advanced, and more accurate, cognitive strategies. This provides increased cognitive accuracy for making more rational decisions and adaptations in a variety of situations. The overall efficacy of this approach probably accounts for the increasing success and acceptance of rational emotive and cognitive behavioral therapies (Ellis & Harper, 1997, pp. vii-xiv; Beck, 1976, p. 4; Wright, 2004, pp. xv-xx). The corrective effect of the edit function on cognitive accuracy greatly enhances the probability that the higher-level objective executive functioning of the DLPFC will have the most rational last word in decision-making (D'Esposito & Postle, 2002, p. 177).

The frontal lobes use stored processing rules for executive functioning, evaluating and prioritizing the available information to determine the best choices for reaching our goals. Declarative memory (i.e. information about how the world works, both *subjectively* and *objectively*) generally appears to be processed for storage by limbic system components, including the hippocampus (Duvernoy, 2005, p. 28), which has significant connectivity with the amygdala (Mesulam, 2000, pp. 2, 58-63). The hippocampus assists in encoding memories with place preference and the amygdala with the assignment of emotional valence (LeDoux, 1996). This information includes semantic and grammatical memory, episodic memory, verbal memory, and biographical memory, along with the rigid rules of our cultural belief systems. Retrieved memories tend to return with the emotions, or limbic valences, that accompanied their initial storage (Phelps & LaBar, 2006, p. 430), albeit modified to some extent by our present emotional state and by time-related decay. This allows past learning to condition present experience and explains to some extent how our cultural belief systems rule us so tenaciously and with such strong emotions. Fortunately, we can ameliorate these subjective emotional reactions by evaluating and editing inaccurate grammar and cultural beliefs, including our faulty assumptions about the cause-effect relationship between thoughts and feelings (Ochsner, 2007, p. 107; 2005, p. 253).

Cognitive Processing

Words and grammar enable us to describe environmental cues, making maximum use of neocortical associations. The level of complexity relates to increased working memory demands. The ability of the frontal lobes to use working memory optimally depends heavily on the availability and quality of the objective process information received from other cerebral regions (Anderson et al, 2002, p. 505) as well as accurate environmental perception. The frontal lobes function best with accurate and timely up-to-date information combined with accurate thought processes i.e., accurate data and the most appropriate, flexible software (i.e. how we think). Thinking uses process information for the assessment and integration of our internal learned information with external environmental information in order to regulate our interaction with the environment and maintain a probabilistic model of anticipated events (Raichle, 2006, p. 13). The software enables the frontal lobes to process executive decisions and to help regulate our emotions appropriately and provide overall homeostasis (Malloy & Richardson, 2001, p. 128).

Grammatical process memory (i.e., acquired and developed rules of thinking) directly biases not only how we process internally stored information but also how we perceive environmental information, cues and stimuli (Decety, 2007, p. 284). Learned, rigid, *dichotomous* process rules preempt flexible thought processes and distort the accuracy of object identification and evaluation. As we might expect, *rigid*, inaccurate, irrational process information leads to inaccurate, irrational information processing, and this in turn yields inaccurately biased executive decisions. Faulty, inaccurate, rigid, dichotomous software bias tends to yield subjectively biased, inaccurate choices and outcomes, which thwart the *flexibility* offered by higher-order, multivariate executive functioning—completing the cycle by reinforcing rigid cultural subjectivity. "No complex system can succeed without an effective executive mechanism, 'frontal lobes.' But the frontal lobes operate best as part of a highly distributed, interactive structure with much autonomy and many degrees of freedom" (Goldberg, 2002, p. 230).

Human Brain and Cognitive Development

Where do inaccurate, irrational thought processes come from? How could we have learned our culturally inherited, faulty, inaccurate, irrational thinking without realizing it? As humans evolved higher-level brain structures, they underwent biological changes that freed them from the constraints of simplistic, concrete language processes and offered the potential for complex, abstract, associative reasoning, made possible by the dramatic evolution of the neocortex and frontal lobes. The capacity for improved frontal lobe reasoning evolved in humans alongside our development of semantics, grammar, and language skills. This same evolution from concrete thinking to the potential for abstraction occurs in an individual's frontal lobe development and connectivity as they grow from childhood to adulthood. Concrete thought processes evolve into more flexible, associative, abstract cognitive capabilities. During these developmental stages—towards puberty and into young adulthood—the frontal lobes continue to mature, with increased connectivity and myelination, especially in the DLPFC (Dennis, 2006, p. 135). This development—along with language, learning, and education—increases the potential maturity of our decision-making skills (i.e. executive functioning) and heightens our ability for cognitive awareness. The frontal lobes play a critical role in this process (Stuss et al, 2001, p. 108).

An episodic increase in cortical pruning takes place in young adolescents around 11 to 14 years of age. This pruning appears to select out the least used or weakest neural connections to make room for the more focused information processing required for improved complex problem solving, social and sexual maturation. According to Hebbian principles (Hebb, 1949), the most used or strongest connections are preserved. Unfortunately, the strongest and most used connections in young humans tend represent the belief systems and associated rigid dichotomous grammatical processing information of their childhood culture, which may become even more firmly embedded in the memory storage of the young brain during this period. The timing of these changes may possibly serve to enhance the resiliency and persistence of these ingrained belief systems.

Cultural Inheritance

As we grow into adulthood, neural efficiency and connectivity to the frontal lobes increases, along with the increasing use of acquired information and grammatical processing rules to make executive decisions. We tend to use our acquired information as if it were factual (Jones, 1998, p. 46), automatically following rigid, dichotomous processing rules that operate outside our awareness. We erroneously tend to identify the first thoughts that come to mind as "facts" and then execute heuristic (arbitrary rules of thumb) justifications and prefabrications to explain away errors based on these "facts." We implicitly assume good-bad relationships between words and reward contingencies learned during our personal developmental history. This may possibly contribute to locking in our cultural belief systems at the expense of more effective, higher-level executive functioning. We seem to favor irrational automatic rules over the acquisition and application of more accurate and up-to-date information. Our decision-making process tends toward a subjective, bottom-up frame of reference with our own personal cultural bias, driven by information filtered through our subjective personal historical matrix, and distorted by the inevitable, implicit, grammatical inaccuracies embedded there.

Our cultural bias not only reflects the ineffective cognitive habits and grammatical inaccuracies of our ancestors, it also represents the rigid, culturally biased misperceptions we inadvertently inherit from generations of unscientific, misinformed, and most often poorly educated elders. Our underlying beliefs derive in large part from their ritualized concrete thinking, faulty assumptions, coincidences imagined as cause and effect , superstitions, myths, magical thinking, etc. (Benjafield, 2007, pp. 321, 333-4). Because children have little experience with which to evaluate different concepts, they uncritically absorb whatever they encounter without regard for its rational usefulness or accuracy. They implicitly and unconditionally absorb rules and information expressed as *truths* without the full benefit of mature executive functions. They tend not to question the factual basis of information, the logic of assumptions, or the reasonableness of conclusions, nor do they have the acquired grammatical framework to do so (LeDoux, 2002, p. 96). This results in a subjective discriminatory bias. In a sense, children are "imprinted," (Lorenz, K. 1965) blind to other possibilities (Benjafield, 2007, p. 267; Shilpa et al, 2007, p. 53).

Information and Process Bias

Concrete learning from early developmental stages tends to accumulate in memory-storage areas, along with the misinformation absorbed from cultural belief systems. The stored information usually reflects the dichotomous hierarchal processes of the parent-to-child interactions under which they formed. If their cultural belief systems remain unchallenged, adults usually exhibit parent-to-child *authoritarian* grammar characteristics in their thinking and interactions, with irrational, prescriptive and imperative demands such as *should*, *must*, *have to*, *need to*, etc. These parental, subjective, dichotomous demands typically "trump" multivariate choices and impede adults from taking responsibility for objective choice-consequence decisions. Even though the parent-to-child interaction style appears to have evolved to beneficially manage and control the child's behavior until the frontal lobes develop, it unfortunately results in implicitly carrying irrational, dichotomous, authoritarian, parental thought and speech habits into adulthood.

Uncertainty and Deviation from the Mean

This irrational thinking usually passes down through many generations and carries into adulthood in part because of our tendency to gravitate toward the familiar (Benjafield, 2007, p. 386) and away from the unfamiliar. Humans have a tendency to experience rewards, or gratification, for sticking with the familiar, and punishment, or anxiety, for venturing into the unfamiliar. From an evolutionary view, this makes a lot of sense, because the unfamiliar may represent a potential danger, while the familiar has at least theoretically been evaluated and found to be safe (Keller & Chasiotis, 2006, p. 277). Familiar events or those with outcomes we deem certain deviate very little from the mean, and as such, generally receive a positive valence with a higher *probability* of favorable outcome.

Uncertain events can deviate significantly from the mean and usually receive a negative valence and a lower favorable probability value. This tends to give certainty the upper hand over the perceived ambiguity and unfavorable probability of uncertainty. This creates a problem for humans determined to cling to illusions of certainty in an uncertain world. However, irrational thinking and heuristics, commonplace rules absorbed from the dominant culture, magically repair these deviations from reality. We use heuristics that rely on faulty grammar and faulty logic to whitewash our errors and inaccuracies and to temper the inherent increase in error variance. This allows humans to alter their perception of reality to explain away any variance by clinging to the irrational *certainty* of their faulty beliefs. Unfortunately, the erroneous irrational *perception* that variance has decreased cannot mitigate the *actual* increase in real error variance and outcome imprecision.

In a sense, the familiar represents comfort, and we human creatures favor our comfortable habits. Humans so favor the familiar that they will often tolerate a great deal of discomfort to hang on to it, sometimes enduring even catastrophe before they will consider the alternative of change, or deviation from the mean. Unfortunately, humans usually tend to err on the side of favoring the familiar and generally receive no logical framework or training to help us develop objective, reasonable, innovative methods for seeking rational alternatives. Indeed, many cultures actually punish attempts to promote cognitive accuracy, some very severely, because it challenges authority and cultural beliefs. This usually prejudices us toward choosing information based on habit, familiarity, and subordination rather than rational utility or applicability, regardless of what might represent our best interest from a more objective perspective.

Authoritarian Communication

Due to the parent-child environment in which they learned these concepts, most adults implicitly continue to use the familiar parental cognitive process in their adult interactions. This hierarchy of communication is authoritarian, vertical, and one-way (parent-to-child) rather than cooperative, horizontal, collaborative, and reciprocal (adult-to-adult, human-to-human). The familiar irrational habits from our past usually preempt more reasonable adult-to-adult communication and block accurate associative reasoning processes, resulting in irrational thought, decision-making, and behavior. Parent-to-child-oriented irrational thought processes generally inhibit reasonable adult-to-adult communication, and this inhibition applies internally when we evaluate our own thoughts and emotions, as well as externally when we relate to others.

The strong tendency of humans to affiliate with groups and gravitate towards the familiar appears to impede adaptive change in humans and cultures, pulling the entire group backward toward more primitive, inaccurate, concrete ways of thinking, decision making, and behaving (Goetz & Shackelford, 2007, p. 13). The cultural inaccuracies we learn often contain an accumulation of past superstitious beliefs that dictate our behavior and soothe the angst of our limbic system (i.e. our emotions). Most cultural information results from implicit grammar and accumulates without the benefit of a scientific framework. Because we grow up under its influence, we generally fail to acquire the critical skills or awareness to assess the validity of our knowledge. Our innate affiliative behavior and inherited rigid thinking, coupled with inaccurate information, unfounded assumptions and lack of awareness, hinders our progress towards maximizing accurate rational thought and behavior, and more harmonious relationships.

Event-level Accuracy

Furthermore, it sets us up to "automatically" use outdated, ritualistic, familiar information and processes for solving a problem at a given moment, excluding new information and shifting us toward the past. Of course, old knowledge often has benefits—but if, by using it, we exclude new and possibly pertinent objective information, our familiar solutions may fail to adequately resolve problems we face in the present or future. We operate in the *now* as if it were identical to the *past*, creating a cognitive time warp. This rigid event-level distortion contributes to our irrational bias, magnified by our rigid, irrational, grammatical software.

Conversely, by selecting the most pertinent, accurate, and current information available and then processing it with flexible, accurate software, we shift our event-level orientation to the present, facilitating the most accurate, best-choice best-outcome decisions. More specifically, we then may use our frontal lobe executive function and working memory to make the most accurate rational choices by flexibly using all available pertinent objective information. In this way, we execute decisions in the present and develop plans for the future with a higher degree of precision and rational probability. This seems preferable to making irrational choices using irrelevant imprecise subjective information from the past, and then using the frontal lobes to retroactively justify and rationalize the decision. Accurate integration of time and space remains critical for problem solving and finding the best solution at any given time (Fuster, 2003, p. 62, 109).

Accurate Evaluations: Information and Processing

We would generally expect to achieve the best outcomes if the frontal lobes use all of the available timely and pertinent information to make decisions, not merely relying on what we learned in the past. This applies especially when we use higher-order executive cognition

(Diamond, 2002, p. 494-95; Braver & Ruge, 2006, p. 307) to evaluate and integrate the emotional components of lower-order limbic cognition to differentiate between choices that feel good but may not serve our best interest, and choices that feel bad but may serve our best interest. Without accurate information about the situation at hand, we might decide on a course of action simply because it feels good or promises familiar rewards, or because it steers us clear of unknown or imagined threats. Furthermore, thinking accurately and rationally seems more likely to yield the best results when dealing with strong emotions, such as when interacting in bonded intimate relationships and affiliations, participating in territoriality and competition for resources, or when memories come with a particular emotion from the past that may represent an irrelevant concern of the present (Heilman, 1997, p. 135). Conversely, research shows that hormones (CRF) generated under stress tend to attenuate higher-order prefrontal cortex functioning, possibly shifting our thought processes towards lower level limbic and automatic implicit cognition. Therefore, we would probably benefit from applying the knowledge that rational higher-order cognitive functioning can preempt or minimize emotional stress and make a significant difference in outcomes. Cognitive therapies focus on rationally remodeling the grammar of belief systems, including percepts, concepts, and processing, to enhance objective awareness, teach stress management and coping skills, help to prevent or relieve depression and anxiety, and improve relationships (Beck, 1976, p. 328).

Benefits of Critical Thinking to Cognitive Accuracy

Critical thinking skills offer potential benefit for overcoming the disadvantages of normative uncritical thinking. Critical thinking encourages the adoption of more objective and scientific guiding principles useful for scrutiny of semantics, grammar, language, culture, assumptions, cause-effect conclusions, and conflict resolution. Critical thinking actually insists on questioning the totality of the argument, including the coherence of the argument process itself by emphasizing accurate, reliable and transparent evidence from a credible source, awareness of motivations and biases, clarity of observation and expression, and reliability of premises and assumptions, inferences, and conclusions. With critical thinking tools, we can evaluate the tactics used for irrational arguments by looking out for false beliefs, distractions, insults, catastrophe, perfect solutions, equivocation, appeal to popularity, authority, or emotions, distortions, false dilemmas, wishful-thinking, explaining by naming, hasty or glittering generalities, irrelevant topics, begging the question, etc.,(Kida, 2006; McInerny, 2005; Jones, 1998; Fisher, 2001; Aubyn, 1957).

Cultural Belief System and Constraints on Accurate Thinking

Current theory of normal human brain functioning indicates that we have the capacity to choose how and what we think, and we have a capacity to adapt to new situations by making better choices (Diamond, 2002, p. 494). By choosing objective information that is more accurate, and flexibly processing it, we increase our chances for determining the most accurate, reasonable, and timely solutions. In other words, we have the potential to bias our choices and outcomes flexibly, in a more accurate, objective way, incorporating semantic and information accuracy, accurate information processing, accurate time-appropriate and context appropriate information. Given this, why would we not choose an accurate rational bias with the highest degree of flexibility? Unfortunately, the subjective characteristics of our implicit grammar, along with personal, social, and cultural biases, impede us from understanding and using the very information that could enable such a choice (Beck, 1979, p. 13). Dichotomous grammar rigidly provides a potential fail-safe mechanism for perpetuating irrational implicitly learned cultural belief systems.

Because of these inherited belief systems, many humans tend to think they *must* be perfect, unblemished and without flaws, that they *must not* be flawed and fallible. A blemish or a mistake means they *are* no good, unworthy, or deserving of punishment. This absolute rating and labeling blocks recognition of personal fallibility, causes poor acceptance of others as fallible human beings, and promotes unscientific, culturally biased, arbitrary category classifications. This leads to rigid, judgmental, and dogmatic cultural belief systems, bigotry, stereotyping, and blind trust, especially blind trust in perceived authority. It also promotes vertical hierarchies with authoritarian, parent-to-child, one-way communication.

We often state our opinions as rigid, true-or-false, absolute statements about the universe, branding them as factual and either right or wrong. We use culturally determined, rigid, dichotomous, authoritarian, imperative, either-or terms, such as *should, must, have to*, and *need to*, implying that we have no other choices or that we *are obligated* to a certain choice. This cultural binding of *shoulds* tends to reinforce simplistic true-false perceptions, rigidity, and predetermination. Rigid, inaccurate, faulty beliefs and assumptions about the cause-effect relationship between thoughts, emotions, and behaviors inaccurately absolve us of responsibility for our own individual evaluations and decisions. Inaccurate definitions and rigid word use, combined with faulty cultural classifications, assumptions, generalities and absolutes, impede reasonable thought processes, hinder personal responsibility, and undermine accurate communication.

Grammatically constraining choices by defining them with rigid words and absolute concepts limits our cognitive options, especially in a *dynamic* world with many variables in a state of frequent change. We commit both a mathematical and a practical error when we arbitrarily restrict the use of multiple, possibly pertinent, variables and rigidly subscribe to culturally determined, dichotomous, *true or false* values when calculating choices and their outcomes (Langer, 2000; Langer & Piper, 1987; McInerny, 2005, p. 94-5). Even though our final choice may be binary, starting with the assumption of only prescriptive, non-contextual, binary choices imposes needlessly strict constraints. Rigidly held, inaccurate cultural belief systems decrease flexibility in our thinking, and limit our ability to develop harmonious relationships both as an individual and with others. These inaccuracies also perpetuate mind-brain dualism, a worldview that artificially separates the *mind* from the *brain* and fosters dichotomous, either-or thinking. Mind-brain dualism dates back to Plato and Descartes, and persists in many subsequent theories and beliefs (Morris & Dolan, 2004, p. 365). A recent study shows this fundamentally inaccurate worldview continues to thrive in the general population, ironically even among psychiatrists, psychologists, and mental health professionals (Miresco & Kirmayer, 2006).

Dichotomous bias, and the accompanying cognitive rigidity it fosters, exhibits many of the same features as dysexecutive functioning — inability to notice, integrate, or appropriately edit internal-external discrepancies in error detection and error correction (Braver & Ruge, 2006, p. 321). Rigid bias creates and perpetuates "reflexive" responses, similar to innate instinctive behaviors, with inflexible stimulus-response paradigms impervious to modification by context or experience (Mesulam, 2002 pp. 14-15, 22-26). Frontal lobe injury often interferes with divergent thinking, perhaps similar to the results one might expect with rigid cognition. Dysexecutive disorders fundamentally limit insight and bias cognition toward the use of primitive dichotomous grammar, inflexible behavior, limited learning from experience, perseveration, and concreteness (Milner & Petrides, 1984), appearing similar to primitive, poorly educated, fundamental cultures and belief systems. Evaluation of diagnostic categories might also benefit from a consideration of objective cognitive accuracy, given the apparent prevalence of dysexecutive features associated with many diagnostic categories. Disorders attributed to personality characteristics frequently present with dysexecutive components and such patients often have a history of developmental adversity and possibly other discrete frontal lobe dysfunction as well.

In addition, substance abuse, anxiety, adjustment disorders, marital discord, obesity and eating disorders may demonstrate unacknowledged frontal lobe dysfunction, possibly due to subjective classifications and biased, subjective, culture-bound heuristic evaluations and treatments (i.e., "we have always done it this way.") For example, consider a culturally biased definition of delusions as "false beliefs based on incorrect inference about external reality and firmly sustained in spite of the *opinion of others* or contrary evidence" (DSM). Evaluating one person's sanity by comparing it to the average opinions of others hardly sounds objective. How would one separate a delusion from prevalent subjective cultural beliefs or widespread superstitions using this definition? A dictionary definition seems a bit more clear, "a false belief or opinion resistant to reason or confrontation with actual fact" (Random House, 1997); however, this still leaves no easy way to distinguish delusions from opinions that differ from a culturally subjective, but factually invalid, consensus opinion. It seems that what many people call *facts*, i.e., beliefs and subjective knowledge, actually serve as subjective grammatical reference points that vary from culture to culture without regard for objective accuracy. How do we distinguish the "cultural facts", i.e., the imagined certainties of static subjective beliefs, from observed, uncertain, probabilistic, dynamic, scientific facts?

Reference Points for Cognitive Accuracy and Rational Bias

Life unfolds as a series of choices and outcomes. We would like to predict the outcome of a particular choice with some degree of certainty, but doing so depends on understanding the many variables in our ever-changing, *dynamic* world. It seems important in this complicated dynamic environment, to accurately assess probable outcomes using flexible, multivariate thought processing and to adequately evaluate our many choices and desired consequences, increasing our contextual response to a given situation. This enhances our overall adaptability. "Choices can be rational or they can be the outcome of irrational processes" (Benjafield, 2007, p.4). A deliberate bias towards rationality tends to enhance our overall accuracy. A rational dynamic bias favors more effective decision-making and increases the probability of more reasonable outcomes. On the other hand, an irrational static bias tends to decrease overall accuracy, leading to irrational decision-making with fewer reasonable choices and fewer objective outcomes. The standards for measuring accurate and rational cognitive bias arise in part from the following assumptions:

- *Acceptance of human imperfection enhances information and process accuracy.* We can accurately characterize humans as flawed and fallible. Accepting our own flaws and fallibilities encourages us to accept others as human beings, albeit with flaws. This acceptance promotes horizontal, human-human, adult-adult collateral communication. It also reduces inaccurate, absolute, dichotomous, or culturally biased classifications, ratings, and labeling, because we do not believe that any person is *all* bad or *all* good. It seems much more reasonable to rate the behavior rather than the person. Rational human acceptance minimizes inaccurate judgmental categories and cultural bigotry while promoting realistic scientific belief systems and healthy skepticism.
- *Flexibility enhances information and process accuracy.* Flexibility generally works better than rigidity for the most accurate planning, problem solving, and compromising in a dynamic world. Rigid, dichotomous, culturally determined terms like *should, must, have to, got to*, and *need to*, restrict options and diversity, while multivariate, preferential terms, such as *I would prefer*, *I would rather*, and *to me, this seems best*, multiply the diversity of possible choices and acceptable rational outcomes. Opinions and preferences replace absolute declarations of right and wrong. Generalities stated as assumptions or deductions represent a higher level of accuracy than generalities misrepresented as true or false *facts* when they actually represent subjective *beliefs*. Such generalizations amplify inaccuracies. Moreover, we improve the precision of our thought processes, decisions, and communications by using the most accurate word definitions, making specific rather than vague statements, using multivariate spectra or gradients for evaluations and revaluations, and avoiding faulty cause-effect conclusions (Browne & Keeley, 2007, p. 147).
- *Awareness of the relationship between thoughts and emotions enhances information, process, and event-level accuracy.* Normally, cognition has a significant causal relationship with our feelings, and realizing this enhances accurate assessment of individual responsibility for thoughts, feelings, and behaviors. Our thoughts, or the rules behind our thoughts, cause or significantly influence our feelings whether we recognize the connection or not. Awareness of this relationship enables us to choose the healthiest and most rational thoughts in order to maximize our emotional and behavioral balance at a given time. Although we might initially react to the situation itself, we largely generate and sustain our emotional reactions to events by what we think or "believe" about them (Ochsner, 2006, p. 245-50). We tend to sustain the emotion long afterwards through the action of implicit internal rules and appraisal habits that affect us almost continuously, generally without our awareness or deliberate direction. As Epictetus wrote in the *Enchiridion* in the first century, "People are disturbed not by things, but by the views which they take of them" (Ellis & Harper, 1997, p. 39).

Cognitive Awareness

Becoming aware of our internal narratives, rules and beliefs about the world and events gives us some understanding of the effect we have on our own individual state and, with that understanding, the ability to objectively moderate our reactions. When something unexpected happens, we often tell ourselves inaccurate, irrational, and overly negative things about the situation, needlessly upsetting and stressing ourselves about it. If instead we insightfully choose to describe the situation as accurately as possible, we can respond with the most appropriate behavior and most reasonable emotion. "Why *must* you upset your self?" (Albert Ellis, attributed). We do not *have to* upset ourselves needlessly or act irresponsibly on irrational thoughts. We improve our individual accountability when we take responsibility for our thoughts and how those thoughts affect our feelings and behaviors. We each have responsibility for managing our own cognitive accuracy, including our individual responsibility for our thoughts, emotions, and behaviors. This enhances reasonable behavior and harmony between humans (Gemba, 2002, p. 146).

Emotional disturbance, in sum, usually stems from your Irrational Beliefs. You can uncover the basic unrealistic ideas with which you disturb yourself; see clearly how misleading these ideas are; and, on the basis of better information and clearer thinking, *change* the Beliefs behind your disturbance (Ellis & Harper, 1997, p. 69).

An orientation towards cognitive accuracy encourages active, objective thinking in the present with proactive, forward-looking, active, event-level evaluations; adaptability; continuous quality improvement; and positive reinforcement and recognition of the importance of rational thought processes. The complexity of life magnifies the importance of thinking accurately. Acceptance of human imperfection, bias towards flexibility, and individual responsibility enhance the accuracy and quality of our cognition and increase the probability of achieving the most reasonable (i.e. *best*) outcomes. This improved quality of accurate thought promotes more harmonious interactions within and between individuals. It also promotes awareness of the benefits of thinking and acting rationally, completing the circle:

It is not the strongest of the species that survive, nor the most intelligent, but the one most responsive to change (Commonly attributed to Charles Darwin, biologist).

Correcting Irrational Biases

However, how can we learn these new skills if, throughout our development, we see few examples of critical thinking and adults fail to teach us to think more logically? How do we improve our cognitive accuracy when we receive little or no foundation for rational or logical information templates in our memory storage areas to build on? Finally, how can we use our acquired, almost "fail safe," irrational, rigid, dichotomous thought processes to learn how to think rationally and make choices that are more rational?

Fortunately, we have the capacity to replace inaccurate, irrational thinking habits with newer, more accurate, rational associative thought processes. It takes effort and practice to learn and use new concepts and retool our brain library and dictionary (Lieberman, 2006, p. 201) with new information and software. The greater the effort (i.e. the more often and harder we practice), the sooner we can replace our habitual, inaccurate, irrational thinking with the new objective skill of accurate rational thinking. This objective shift in the way we think uses higher-level, multivariate, associative reasoning instead of rigid, black-and-white, either-or cognition to evaluate choices and outcomes. Rational associative reasoning tends to maximize accurate executive cognition and decision-making. This increased accuracy contributes directly to thinking more rationally with more reasonable outcomes, because our higher-level executive functioning has the utility of the "last word."

Implications of Irrational Bias

> "Our present problems cannot be solved at the level of thinking at which they were created" (Albert Einstein, attributed).

When applied to decision-making, rigid, irrational biases tend to employ a floating reference point; in contrast, biases that are more flexible and rational tend to operate from a stable reference point. How can we explain this seeming paradox? Understanding it depends on the perspective of the observer. An observer using the same irrational biases, grammar, terms, and definitions used to construct the irrational reference point will find this conundrum baffling. As previously stated, humans rigidly cling to their cultural belief systems, no matter how subjectively antiquated or irrelevant. These belief systems take hold during early learning, along with the related dominance characteristics of parent-child interactions and obedience to authority (Milgram, 2004, pp.114-5, 136-47). We learn, whether directly or indirectly, that *good* is rewarded and *bad* is punished, that *right* is rewarded and *wrong* is punished, and that parents know what is *right* and what is *wrong*. The problem arises when we apply these static dichotomous grammatical constructs to the complexity of the dynamic variables of the world we live in. Humans tend to perceive the rules for decision making in hierarchal, dichotomous and absolutistic terms: is, is not, black, white, should, must, have to, ought to, etc. *You must do that or you will be punished*. This creates a problem when the must of today becomes the must not of tomorrow because some variable or context has changed.

Reference Point Drift

This dichotomous grammar then leads to retrospective judging, faultfinding, blaming, and punishing. *You should not have done that*! This rationale might work if we had a means of predicting the future in our dynamic environment; given that we do not, we look for another mechanism to explain the unwelcome outcome. To resolve the apparent discrepancies with reality, we simply move our reference point: We redefine success or modify our recollection of errors to cast the current outcome in a better light (Schlacter, 2001, p. 151). By moving the reference point, we can always be right, always be above average, and never make a mistake. We simply move the reference point to under-value others or over-value ourselves, and this creates the irrational but comforting experience of "self-esteem." We perhaps learn this skill in childhood when governed by the multiple, subjective, and inconsistent reference points of parents, teachers, neighbors, friends, celebrities, etc.

We also learn to easily construct subjectively biased, over-generalized groups and classes and use them to denigrate and subjugate others at will. We can easily over-generalize or under-define representative classifications and prove most any point using imprecise labels. We can round up or down at will. We have our choice of measuring instruments and usually tend to choose the yardstick that casts us in the most favorable light, with little thought about accuracy. Since each individual chooses their own rules from their own frame of reference, we can use one rule to justify and a different rule to vilify any one experience to suit our own beliefs and goals (Kida, 2006, p. 78), thereby potentially artificially elevating self-esteem. This works well for confirmation bias (Kida, 2006, p. 160).

Social cognitive neuroscience has demonstrated that when humans see others as humans, they tend to moderate their prejudices (Mitchell et al, 2006) and that positive experiences with others can erase learned cultural stereotypes (Phelps & LaBar, 2006, p. 440). Milligan found that when volunteer experimenters, under commands of authority, viewed experimental subjects as "unworthy persons" rather than humans, their willingness to punish them increased (Milligan, 2004, p. 161). In another study, volunteer experimenters punished subjects less when the experimenter was in closer proximity to subjects (Milligan, 2004, pp. 37). This suggests that subjective human labeling contributes to prejudice and punishment by making others less than human. When we remove the subjective labels and see others objectively as humans, like us, we can acknowledge that we belong to the same class, *Homo sapiens*. We may not approve of certain behaviors in others but we have the option of labeling the behavior rather than the person.

Humans often blame others for their own thoughts, emotions, and behaviors. Since we fear being wrong, we simply decide that someone else *made* us think, feel, or behave a certain way. We shift the blame to a scapegoat. They are at fault. They are to blame. Simply by adjusting the reference point, we easily pass along the blame and judge ourselves not guilty. This approach may succeed in bolstering the "self-esteem" of the blamer, but usually at the expense of anger or guilt of the blamed, due to their resentment or internally demoted *self-esteem*. This can have disastrous effects on long-term relationships.

Irrationality of Self-Esteem versus Rational Acceptance of Human Imperfection

Unfortunately, self-esteem represents an authoritative artifact of irrational cultural constructs that depend on dichotomous judgments of right and wrong, perfect and imperfect, and other culturally defined, rigid, subordinate subjective values (Milligan, 2004, p. 147). A stable reference point based on cognitive accuracy uses a consistent spectrum to gauge behavioral achievement, performance, and error, in effect giving us standards that provide a more rational reality for testing outcomes, while avoiding statements about individual worth. With a rational bias, we accept that a certain rate of failure is inevitable; this minimizes the motivation to "fix" our mistakes by justifying, rationalizing, or blaming others. This, in turn, helps us resist the temptation to float the mean or tweak the reference point to make our decisions appear more above average or always correct. We can freely admit our error and concentrate on improvement.

Acceptance of human imperfection tends to be a more objective and workable construct than self-esteem (Ellis, 2005) because it starts with reality, the rational assertion that human beings have flaws and make fallible decisions, and therefore err on occasion. If we embrace human fallibility and the uncertainty of our choices, we do not suffer an intolerable shock when we fail, and we do not find it necessary to change history or redefine our success and achievements to feel happy. However, even if we rationally accept the premise that we will rarely predict the future with 100% accuracy, we still face a potential challenge from the parts of our brain that store and apply our cultural beliefs and irrational biases. As the gatekeeper of the limbic system, in charge of safety, the amygdala appears to continue to preferentially reward the perceived certainty of familiar choices to such an extent that we still routinely choose against our own best interests. We even chisel our principles and beliefs in stone and refuse to admit to any degree of error. This mollifies the amygdala, when operating on irrational beliefs with black and white rigidity in a mostly gray world. Then, instead of rationally evaluating and predicting the future, we habitually use our exceptional frontal lobes to *rationalize the past*. We make up after-the-fact explanations and excuses to *justify* the errors of our behaviors, refute reality, or recast history according to our own wishes while magically preserving our self-esteem. Fortunately, the amygdala responds to cognitive control (Phelps, 2004, p. 1013). Cognitive and rational therapies temper the irrational anxiety of the amygdala by providing education in objectivity and competent rational strategies.

Our human tendency to rationalize and justify may come from our evolutionary roots. As our primitive semantics, grammar, and language arose from symbolic grunts, gestures, and facial expressions to form the earliest beginnings of more formal communication, we began to use words to label objects and actions (rock...not rock, bad...not bad). From this, dichotomous grammar evolved. This may have worked well in primitive cultures to improve group adaptation to the environment, using dominance, punishment, reward for accomplishments, and faulty interpretation of coincidences extrapolated to causation. The *veridical* nature of black and white, dichotomous grammar increases the probability of error in the mostly gray and ever-changing *dynamic* world we live in. Ironically, these increased errors of dichotomous cognition also increase deviation from the mean, along with increased variability in dichotomous rule adherence. These errors and deviations cause internal-external evaluation discordance. The easiest irrational corrective solution involves generating explanations for these dissonant discrepancies in the form of irrational beliefs, heuristics, and their accompanying biases.

In early primitive cultures that lacked scientific knowledge, language arose out of reasonable attempts to label objects and understand relationships between humans and the environment initially using nouns (objects) and verbs (action). Since nouns, verbs, and language predated writing, we might reasonably assume they sustained an early narrative historical record and that progress in objective knowledge made it increasingly more difficult to explain away uneducated, unscientific beliefs, faulty labeling, and transfer of fault and blame to others. However, heuristics may have evolved as a form of mental shortcuts that were then extrapolated to justifications for irrational beliefs, thoughts, and behavior. Heuristics represent easily stated and applied "rules of thumb" that develop as a part of most cultural belief systems. While they generally purport to convey compact bits of wisdom, they more often embody cognitive and verbal tools for evading reality. They enable humans to correct errors by explaining them away, by generalizing, justifying, dignifying, signifying, rationalizing, blaming, bullying, revising, labeling, etc. We can then resort to the prefabrication of even more *preposterous* irrational rules somewhat akin to confabulation.

With irrational, dichotomous grammar, humans can create a scorecard, noting their own successes and ignoring their own errors, while tracking the errors of others and ignoring their successes. This allows us to maintain a perception of "self-esteem," and remain above the mean, above average and perfect. Heuristics and irrational beliefs make it easier to evaluate internal-external errors and error corrections from our own frame of reference. Unfortunately, dichotomous grammar increases errors and decreases error detection, whether we can explain them away or not. Objective multivariate grammar generally makes more efficient use of the limited storage space afforded to humans (Nowak, 2006, pp. 249-86) than irrational grammar can. Rational multivariate grammar seems to offer increased accuracy without the convoluted perturbations usually required to correct for errors after the fact under subjective dichotomous grammar. It would seem plausible then that rational, multivariate grammar and acceptance of human imperfection offers a more efficient and a more reasonable construct than irrational grammar and self-esteem.

Discussion, Conclusions, and Considerations

We have the resources to counter these mostly culturally determined irrational and inaccurate grammatical habits by diligent pursuit of accurate cognition, even in the face of doubt and uncertainty. We can retrain the brain to value rational, probabilistic outcomes, accept human imperfection in place of self-esteem, and use accurate information and processing to make decisions about current situations. Each of these practices can eventually enable us to use the processing capabilities of our new brain to direct the older, survival-oriented brain structures (MacLean, 1990; Milgram, 2004) toward more rational pursuits. We can then focus on achieving objective improvement and obtaining better outcomes for the future (Panksepp, 1998, p. 260). This suggests that establishing reference points for cognitive accuracy may provide a naturalistic rational framework for future research that will demonstrate an ecologically valid unifying theory for brain and behavior in *Homo sapiens*. Cognitive accuracy appears to represent an evolution-friendly *multivariate* grammar that might impart greater homeostatic benefits. Objective multivariate grammar meets requirements for simplicity (Occam's razor), and seems to offer a more efficient homeostatic mechanism, possibly minimizing the liabilities incurred by the inherent inaccuracies of subjective dichotomous grammar. Thus, cognitive accuracy may provide useful fundamental principles for establishing a scientific reference point for the evaluation of rational behaviors, choices, and consequences. A potential scientific method based on cognitive accuracy includes three fundamental principles:

- information accuracy: seeking and using objective information based on empirical observation; premise, deduction, conclusions, and testing
- thought process accuracy: making evaluations and decisions flexibly with critical thinking, multivariate terminology, and awareness of individual responsibility
- event-level accuracy: connecting and verifying both information and decisions in a time- and context-dependent manner to increase the relative probability of more accurate predictions of rational outcomes (Bailey, 2007, in press; Bailey, 2006)

Each component helps to ensure reliable cognitive functioning. Event-level accuracy is especially important for maximizing context specificity by accurately integrating cognition (information and information processing), and behavior at a given moment and place. This identifies, acknowledges, and "preserves an independent reality" (Minkowski, 1952, p. 76).

The next shift in human primate cognitive evolution could potentially consist of the implementation and integration of cognitive accuracy into all major areas of scientific study, thereby fostering individual and cultural acceptance and direct application. If so, then thinking more flexibly, accepting uncertainty, and living rationally in the present may become the norm instead of the exception. Increasing our degree of accurate, rational thinking would appear to enhance our prospects for living rationally in the present and planning for a more reasonable and likely future. As accurate, rational thoughts and behaviors increase, irrational thoughts and behaviors tend to decrease. Thinking reasonably and rationally relies on the accuracy of higher-order executive functioning, with the potential to transport humankind to a more human and humane future (Hendelman, 2006, p. 238).

Accurate, rational thinking increases flexibility and tends to maximize appropriate choices, enhancing achievement of preferred outcomes while minimizing undesirable irrational outcomes, without the liabilities of irrational cognition (Fine, 2006, p. 208). Fortunately, we have the ability to replace inaccurate, irrational, parental, absolute thinking by learning and practicing new habits of accurate, flexible, rational, and reasonable logical critical thinking, thus improving overall adaptability. We have these rational tools available, but they often go unrecognized, overlooked, or even belittled (Jones, 1998, pp. 6-7). Ideally, we will achieve this accurate, rational, cognitive evolutionary step before we trigger an irrationally induced catastrophe with our current habits of rigidity, disharmony, self-loathing, and aggression (Beck, 1999).

Science has understood the negative effects of subjective cultural beliefs on the evaluation of objective knowledge for some time. How we measure something in large part defines what we measure. "What you get from a measurement depends on what you choose to measure" (Lindley, 2007, p. 155). We tend to calibrate our personal yardsticks to the culture in which we find ourselves. Such a yardstick measures only what that culture values. Each culture's yardstick shows normal, despite large differences between the beliefs and behaviors of various cultures. This potentially leaves us prey to the accumulated inaccuracies of our forebears.

Cognitive accuracy represents a reference point biased toward objective cognitive accuracy in order to measure human thought and behavior across cultures. This reference point does not change as you go from one culture to the next, theoretically transcending the inaccuracies of cultural belief systems. Cultures may change, but the yardstick remains the same, unless and until scientific advances using cognitive accuracy indicate beneficial objective adjustments. For accurate evaluations, we do best to calibrate our cognitive yardstick with the most accurate, timeliest information, applied consistently and rationally in the present.

This article has reviewed the intimate connection between semantics, grammar, cognition, and the functioning of various brain structures, and posited an equally intimate association between semantics, grammar, cognition, emotion, and behavior. These associations suggest an explanation for the persistence of subjective cultural belief systems and their inherent deficiencies in objectivity, accuracy, and flexibility. Can we accept the premise that objective, timely, accurate processing and accurate information usage might efficiently yield more satisfactory and reasonable results than rigid automatic responses based on inherited imprecise beliefs? If so, it would seem beneficial to society for science to orient new research to incorporate the psychological value of objective relative belief systems, using scientific insights from cognitive neuroscience, anthropology to zoology, ethology to neuropsychology, cognitive behavioral science to neuroeconomics, affective neuroscience, evolutionary dynamics etc., along with the tools of cognitive accuracy.

Cumulative prospect theories and neuroeconomics (Kahneman & Tversky, 1992; Glimcher et al, 2005) suggest that humans weigh the relative probabilities of achieving a particular outcome along an economic gradient of values that influence cognition, emotions, and behavior. Our tendency to seek reward and familiarity and generally avoid risk, uncertainty, and pain, combined with the inertia of our acquired cultural belief systems, biases our value analysis and skews choice-outcome decisions disproportionately in favor of the familiar or the presumed socially desirable outcome (Tom et al, 2007). Moreover, cultural beliefs are resistant to transformation once established (Panksepp, 1998, p. 245; Lane et al, 2000, p. 409), suggesting that whatever adult behavior (including cognition) we wish to promote in a society, we will do best to teach during childhood development and human maturation (Tse et al, 2007). For many generations, parents have raised children to hold the same rigid, unexamined beliefs they hold themselves. If we instead model humane treatment of others as humans, flexibly apply objective critical thinking, and demonstrate reasonable behavior for our children, we have some hope they will instead value cognitive accuracy as adults. Luria, many years ago, found evidence that early education can teach children the ability to use logical thought (Luria, 1981, p. 209). This suggests that exposure to and education in higher-level logical reasoning, throughout development, could enhance rational cognition and behavior.

Once a sufficient number of adults acquire the skills and habit of cognitive accuracy, their interactions with children would likely pass on fewer faulty beliefs and thought processes. Importantly, at an early age, children raised by these adults would have the opportunity to learn to develop and extend their cognitive accuracy. As adults, they would have more skills for applying cognitive accuracy to address and resolve problems with competent critical thinking, emotional balance, and reasonable behaviors. This would enable each successive generation to build on the successes of their parents, rationally living in the present (Korzybski, 1958, p. 231).

Science, by virtue of its central methodology (Browne & Keeley, 2007, p. 119), speaks for the value of accurate, rational, logical objective reasoning and critical thinking, and for advocating the teaching of objective rational cognition, which can potentially guide society towards increasingly humane thought and behavior. Future research on human behavior and neuroscience can contribute to this goal by taking into account the interdependence of biology and society. If we have the choice to conduct research that grounds our cultural belief systems in neuroscience, and it seems we do, while producing more reasonable and responsible adults who use cognitive accuracy and rational evaluations to make decisions, can we afford not to do so?

In the final analysis, we have to depend on our rich resources of rationality to recognize and modify our irrationality...We can recognize that our own interests are best served by applying reason. In this way, we can help to provide a better life for ourselves, others, and the future children of the world, (Beck, 1999, p. 287).

Author Notes

Charles E. Bailey, M.D. is a General Psychiatrist and Clinical Research Psychopharmacologist in Orlando, Florida. cbailey1@cfl.rr.com

Acknowledgments

A special thanks to Nora Miller for her tireless editing efforts; posthumously to Albert Ellis for his editorial comments and encouragement; to Elkhonon Goldberg for his encouragement and sage advice; to M.-Marsel Mesulam for his very thoughtful suggestions; to Khrista Boylan for her valuable input; and to Scott Rauch and Carl Senior for taking time out of their busy schedules to make much appreciated and insightful editorial comments.

References

Adolphs, R. (2006). What is special about social cognition? In J. T. Cacioppo, P. S. Visser, & C. L. Pickett (Eds.), *Social neuroscience, people thinking about people* (pp. 269-285). Cambridge, MA: MIT Press.

Alexander, G. E., Delong, M. R., & Strick, P. L. (1986). Parallel organization of functionally segregated circuits linking basal ganglia and cortex. *Annual Review of Neuroscience*, 9, 357-381.

American Psychiatric Association: *Diagnostic and Statistical Manual of Mental Disorders*, Fourth Edition. Washington, DC: American Psychiatric Association, 1994.

Anderson, V., Levin, H. S. & Jacobs, R. (2002). Executive functions after frontal lobe injury: A developmental perspective. In D. T. Stuss & R. T. Knight (Eds.), *Principles of frontal lobe function* (pp. 504-527). New York, NY: Oxford University Press.

Aubyn, G. St. (1957). *The art of argument*. USA: Emerson Books.

Avrutin, S. (2006). Weak Syntax. In Y. Grodzinsky & K. Amunts (Eds.), *Broca's region* (pp. 49-61). New York, NY: Oxford University Press.

Baddeley, A. (2002). Fractionating the central executive. In D. T. Stuss & R. T. Knight (Eds.), *Principles of frontal lobe function* (pp. 246-259). New York, NY: Oxford University Press.

Bailey, C. E. (2006). A general theory of psychological relativity and cognitive evolution. *ETC: A Review of General Semantics*, 63, 278-289.

Bailey, C. E. (2007) in press. Cognitive accuracy and intelligent executive functioning in the brain and in business. In C. Senior & M. J. R. Butler (Eds.), *The social cognitive neuroscience of business*. New York, NY: Annals of the New York Academy of Sciences.

Barbas, H. (2006). Organization of the principle pathways of prefrontal lateral, medial, and orbitofrontal cortices in primates and implications for their collaborative interaction in executive functions. In J. Risberg & J. Grafman (Eds.), *The frontal lobes, development, function and pathology* (pp. 21-68). New York, NY: Cambridge University Press.

Bartel, A. & Zeki S. (2004). The neural correlates of maternal and romantic love. *Neuroimage*, 21, 1155-1166

Baum, E. B. (2004). *What is thought?* Cambridge, MA: MIT Press.

Beck, A. T. (1999). *Prisoners of hate: The cognitive basis of anger, hostility, and violence.* New York, NY: HarperCollins.

Beck, A. T. (1976). *Cognitive therapy and the emotional disorders*. New York, NY: International University Press.

Benjafield, J. G. (2007). *Cognition* (3rd Ed.). Canada: Oxford University Press.

Bloom, P. (2002). *How children learn the meaning of words.* Cambridge, MA: MIT Press.

Braver, T. S. & Ruge, H. (2006). Functional neuroimaging of executive functions. In R. Cabeza & A. Kingstone (Eds.), *Handbook of functional neuroimaging of cognition* (pp. 307-348). Cambridge, MA: MIT Press.

Browne, M. N. & Keeley, S. M. (2007). *Asking the right questions: A guide to critical thinking* (8th ED.). CITY? NJ: Pearson Prentice Hall

Boyd, R. & Richerson, P. T. (2005). *The origin and evolution of cultures* (p. 206). New York, NY: Oxford University Press.

Broca, P. (1877). Rapport sur un memorie M. Armund de Fleury intitule: De l'inegalite dynamique des deux hemispheres cerebraux. *Bull Academy of Medicine*, 6, 508-539.

Cacioppo, J. T. & Berntson, G. G. (2004). Social neuroscience. In M. S. Gazzaniga (Ed.), *The cognitive neurosciences* III (pp. 977-985). Cambridge, MA: MIT Press.

Caldwell, H. K. & Young III, W. S. (2006). Oxytocin and vasopressin: Genetics and behavioral implications. In R. Lim (Ed.) *Handbook of neurochemistry and molecular neurobiology: Neuroactive proteins and peptides*, 3rd Ed. (pp. 573-607). New York, NY: Springer

Capaldi, N. (1987). *The art of deception: An introduction to critical thinking* (p.17). New York, NY: Prometheus Books.

Cho, M. M., Devries, A. C., Williams, J. R. & Carter, C. S. (1999). The effects of oxytocin and vasopressin on partner preferences in male and female prairie voles. *Behavioral Neuroscience*, 113, 1071-1079.

Chow, T. W. & Cummings, J. L. (2007). Frontal-Subcortical circuits. In B. L. Miller & J. L. Cummings (Eds.), *The human frontal lobes*, 2nd Ed. (pp. 25-43). New York, NY: Guilford Press.

Cummings, J. L., & Miller, B. L. (2007). Conceptual and clinical aspects of the frontal lobes. In B. L. Miller & J. L. Cummings (Eds.), *The human frontal lobes*, 2nd Ed. (pp. 12-21). New York, NY: Guilford Press.

Curtis, C. E. & D'Esposito, M. (2006). Functional neuroimaging of working memory. In R. Cabeza & A. Kingstone (Eds.), *Handbook of functional neuroimaging of cognition* (pp. 269-306). Cambridge, MA: MIT Press.

Davachi, L., Romanski, L. M., Chafee, M. V. & Goldman-Rakic, P. S. (2004). Domain specificity in cognitive systems. In M. S. Gazzaniga (Ed.), *The cognitive neurosciences* III (pp. 665-678). Cambridge, MA: MIT Press.

Damasio, A. R. (2000). A second chance for emotion. In R. D. Lane & L. Nadel (Eds.), *Cognitive neuroscience of emotion* (pp. 12-23). New York, NY: Oxford University Press.

Deardorff, A. V. (2006). *Terms of trade: Glossary of international economics*. Hackensack, NJ: World Scientific Publishing Co.

Decety, J. (2007). A social cognitive neuroscience model of human empathy. In E. Harmon-Jones & P. Winkielman (Eds.), *Social neuroscience: Integrating biological explanations of social behavior* (pp. 246-270). New York, NY: Guilford Press.

Delgado, M. R. (2007). Reward-Related responses in the human striatum. In B. W. Balleine, K. Doya, J. O'Doherty & M. Sakagami (Eds.). Reward and decision making in corticobasal ganglia networks. *Annals of the New York Academy of Sciences*, 1104, 70-88.

Dennis, M. (2006). Prefrontal cortex: Typical and atypical development. In J. Risberg & J. Grafman (Eds.), *The frontal lobes, development, function and pathology* (pp. 128-162). New York, NY: Cambridge University Press.

D'Esposito, M. & Postle, B. R. (2002). The organization of working memory function in lateral prefrontal cortex: Evidence from event-related functional MRI. In D. T. Stuss & R. T. Knight (Eds.), *Principles of frontal lobe function* (pp. 168-187). New York, NY: Oxford University Press.

Diamond, A. (2002). Normal development of prefrontal cortex from birth to young adulthood: Cognitive functions, anatomy, and biochemistry. In D. T. Stuss & R. T. Knight (Eds.), *Principles of frontal lobe function* (pp. 466-503). New York, NY: Oxford University Press.

Duvernoy, H. M. (2005). *The human hippocampus* (3rd Ed.). New York, NY: Springer-Verlag Berlin Heidelberg.

Edelman, G. M. (1992). *Bright air, brilliant fire: On the matter of the mind.* New York, NY: Basic Books.

Ellis, A. (2005). *The myth of self-esteem: How rational emotive behavior therapy can change your life forever.* Amherst, NY: Prometheus Books.

Ellis, A. & Harper R.A. (1997). *A guide to rational living* (3rd Ed.). N. Hollywood, CA: Melvin Powers Wilshire Book Company (Original work published 1976).

Ferris, C. F. (2006). Neuroplasticity and aggression: An interaction between vasopressin and serotonin. In R. J. Nelson (Ed.), *Biology of Aggression* (pp. 163-175). New York, NY: Oxford University Press.

Ferstl, E. C. (2007). The functional neuroanatomy of text comprehension. In F. Schmalhofer & C. A. Perfetti (Eds.), *Higher level language process in the brain: Inference and comprehension processes* (pp. 53-102). Mahwah, NJ: Lawrence Erlbaum Associates Inc.

Fine, C. (2006). *A mind of its own: How your brain distorts and deceives* (p. 208). New York, NY: Norton.

Fisher, A. (2001). *Critical thinking: An introduction.* Cambridge, UK: Cambridge University Press.

Fisher, H. E., Aron, A., Mashek, D., Li, H. & Brown, L. L. (2002). Defining the brain systems of lust, romantic attraction, and attachment. *Archives of Sexual Behavior*, 31, 413-419.

Frith, C., Rees, G., Macaluso, E. & Blakemore, S. (2004). Higher cognitive functions: Mechanisms of attention. In R. S. J. Frackowiak, K. J. Friston, C. D. Frith, R. J. Dolan, C. J. Price, S. Zeki, J. Ashburner & W. Penny (Eds.), *Human Brain Function* (pp. 245 -268) (2nd Ed). San Diego, CA: Elsevier Science USA.

Fuster, J. M. (2003). *Cortex and mind, unifying cognition.* New York, NY: Oxford University Press.

Gazzaley, A. & D'Esposito, M. (2007). Unifying prefrontal cortex function: Executive control, neural networks, and top-down modulation. In B. L. Miller & J. L. Cummings (Eds.), *The human frontal lobes*, 2nd Ed. (pp. 187-206). New York, NY: Guilford Press.

Gemba, H. (2002). Motor programming for hand and vocalizing movements. In D. T. Stuss & R. T. Knight (Eds.), *Principles of frontal lobe function* (pp. 127-148). New York, NY: Oxford University Press.

Giesbrecht, B., Kingstone, A., Handy, T. C., Hopfinger, J. P. & Mangun, G. R. (2006). Functional neuroimaging of attention. In R. Cabeza & A. Kingstone (Eds.), *Handbook of functional neuroimaging of cognition* (pp. 85-111). Cambridge, MA: MIT Press.

Glimcher, P. W., Dorris, M. C. & Bayer, H. M. (2006). Physiological utility theory and the neuroeconomics of choice. *Games and Economic Behavior*, 52, 213-256.

Goetz, G. T. & Shackelford, T. K. (2007). Introduction to evolutionary theory and its modern application to human behavior and cognition. In S. M. Platek, J. P. Keenan & T. K. Shackelford (Eds.), *Evolutionary cognitive neuroscience* (pp. 5-19). Cambridge, MA: MIT Press.

Goldberg, E. (2002). *Executive brain: Frontal lobes and the civilized* mind. New York, NY: Oxford University Press.

Goldman-Rakic, P. S. (1995). Architecture of the prefrontal cortex and central executive. *Annals of the New York Academies of Science*, 769, 71-83.

Grafman, J. (2002). The structured event complex and the human prefrontal cortex. In D. T. Stuss & R. T. Knight (Eds.), *Principles of frontal lobe function* (pp. 292-310). New York, NY: Oxford University Press.

Graybiel, A. M. & Saka, E. (2004). The basal ganglia and the control of action. In M. S. Gazzaniga (Ed.), *The cognitive neurosciences* III (pp. 495-510). Cambridge, MA: MIT Press.

Hauser, M. D., Chomsky, N. & Fitch, W. T. (2002). The faculty of language: What is it, who has it, and how did it evolve? *Science*, 298, 1569-1579.

Hebb, D. O. (1949). *The organization of behavior.* New York, NY: Wiley.

Heilman, K. M. (1997). The neurobiology of emotional experience. In S. Salloway, P. Malloy & J. L. Cummings (Eds.), *The neuropsychiatry of limbic and subcortical disorders* (pp. 133-142). Washington, DC: American Psychiatric Press.

Heimer, L., Alheid, G. F., de Olmos, J. S., Groenewegen, H. J., Haber, S. N., Harlan, R. E. & Zahm, D. S. (1997). The Accumbens: Beyond the core-shell dichotomy. In S. Salloway, P. Malloy & J. L. Cummings (Eds.), *The neuropsychiatry of limbic and subcortical disorders* (pp. 43-70). Washington, DC: American Psychiatric Press.

Hendelman, W. J. (2006). *Atlas of functional neuroanatomy* (2nd Ed.). Boca Raton, FL: CRC Press.

Hooker, C. I. & Knight, R. T. (2006). The role of lateral orbitofrontal cortex in the inhibitory control of emotion. In D. H. Zald & S. L. Rauch (Eds.), *The orbital frontal cortex* (pp. 307-324). New York, NY: Oxford University Press.

Insel, T. R. (1997). A neurobiological basis of social attachment. *American Journal of Psychiatry,* 154, 726-735.

Jones, M. D. (1998). *The thinker's toolkit: 14 powerful techniques for problem solving* (p. 6-7, 14-23, 33, 46). New York, NY: Three Rivers Press.

Kahneman, D., Slovic, P. & Tversky, A. (1982). *Judgment under Uncertainty: Heuristics and Biases.* Cambridge, MA: Cambridge University Press.

Kaufer, D. L. (2007). The dorsolateral and cingulate cortex. In B. L. Miller & J. L. Cummings (Eds.), *The human frontal lobes*, 2nd Ed. (pp. 44-58). New York, NY: Guilford Press.

Keller, H & Chasiotis, A. (2006). Evolutionary perspectives on social engagement. In P. T. Marshall & N. A. Fox (Eds.), *The development of social engagement: Neurobiological perspectives* (pp. 275-303). New York, NY: Oxford University Press.

Kida, T. E. (2006). *Don't believe everything you think: The 6 basic mistakes we make in thinking.* Amherst, NY: Prometheus Books.

Korzybski, A. (1958). *Science and sanity: An introduction to non-aristotelian systems and general semantics* (4th Ed.). Lakeville, CT: International Non-Aristotelian Library Publishing (Original work published 1933).

Langer, E. J. (2000). Mindful learning. *Current Directions in Psychological Science*, 9, 220-223.

Langer, E. J. & Piper, A. J. (1987). The prevention of mindlessness. *Journal of Personality and Social Psychology*, 53, 280-287.

Lane, R. D., Nadel, L. & Kaszniak, A. W. (2000). The future of emotion research from the perspective of cognitive neuroscience. In R. D. Lane & L. Nadel (Eds.), *Cognitive neuroscience of emotion* (pp. 407-410). New York, NY: Oxford University Press.

LeDoux, J. (2002). *Synaptic Self: How our brains become who we are.* New York, NY: Russell Sage Foundation.

LeDoux, J. (1996). *The emotional brain.* New York, NY: Simon & Shuster.

Lee, D. & Seo, H. (2007). Mechanisms of reinforcement learning and decision making in the primate dorsolateral prefrontal cortex. In B. W. Balleine, K. Doya, J. O'Doherty & M. Sakagami (Eds.). Reward and decision making in corticobasal ganglia networks. *Annals of the New York Academy of Sciences*, 1104, 108-122.

Lichter, D. G. & Cummings, J. L. (2001). Introduction and overview. In D. G. Lichter & J. L. Cummings (Eds.), *Frontal-Subcortical circuits in psychiatric and neurological disorders* (1-43). New York, NY: Guilford Press.

Lieberman, P. (2006). *Toward an evolutionary biology of language.* Cambridge, MA: Harvard University Press.

Lim, M. M., Wang, Z., Olazabal, D. E., Ren, X., Terwilliger, E. F. & Young, L. J. (2004). Enhanced partner preference in a promiscuous species by manipulating the expression of a single gene. *Nature*, 429, 754-757.

Lindley, D. (2007). *Uncertainty: Einstein, Heisenberg, Bohr, and the struggle for the soul of science.* New York, NY: Doubleday.

Logothetis, N. K. (2004). Higher cognitive functions. In M. S. Gazzaniga (Ed.), *The cognitive neurosciences* III (pp. 849-969). Cambridge, MA: MIT Press.

Long, D. L., Baynes, K. & Prat, C. (2007). Sentence discourse representation in two cerebral hemispheres. In F. Schmalhofer & C. A. Perfetti (Eds.), *Higher level language process in the brain: Inference and comprehension processes* (pp. 329-353). Mahwah, NJ: Lawrence Erlbaum Associates Inc.

Lorenz, K. (1965). *Evolution and modification of behavior.* Chicago, IL: University of Chicago Press.

Luria, A. R. (1981). *Language and cognition.* J. V. Wertsch (Ed.).Washington, DC: John Wiley & Sons Inc.

Luria, A. R. (1966/1980). *Higher cortical functions in man.* New York, NY: Basic Books.

MacLean, P. D. (1990). *The triune brain in evolution.* New York, NY: Plenum Press.

Malloy, P. F. & Richardson, E. D. (2001). Assessment of frontal lobe functioning. In P. S. Salloway, P. F. Malloy & J. D. Duffy (Eds.), *The frontal lobes and neuropsychiatric illness* (pp. 125-137). Washington, DC: American Psychiatric Publishing.

McInerny, D. Q. (2005). *Being logical: A guide to good thinking.* New York, NY: Random House.

Mega, M. S. & Cummings, J. L. (2005). Frontal subcortical circuits: Anatomy and function. In P. S. Salloway, P. F. Malloy & J. D. Duffy (Eds.), *The frontal lobes and neuropsychiatric illness* (pp. 15-32). Washington, DC: American Psychiatric Publishing.

Mesulam, M.-Marsel (2000). *Principles of behavioral and cognitive neurology* (2nd Ed.). New York, NY: Oxford University Press.

Mesulam, M.-Marsel (2002). The human frontal lobes: Transcending the default mode through contingent coding. In D. T. Stuss & R. T. Knight (Eds.), *Principles of frontal lobe function* (pp. 8-30). New York, NY: Oxford University Press.

Mesulam, M.-Marsel (1985). Patterns in behavioral neuroanatomy: Association areas, the limbic system, and hemispheric specialization. In M. Mesulam (Ed.) *Principles of Behavioral Neurology* (pp. 1-70). Philadelphia, PA: F. A Davis.

Middleton, F. A. & Strick, P. L. (2001). A revised neuroanatomy of frontal-subcortical circuits. In D. G. Lichter & J. L. Cummings (Eds.), *Frontal-Subcortical circuits in psychiatric and neurological disorders* (44-58). New York, NY: Guilford Press.

Milgram, S. (2004). *Obedience to authority* (first published 1974). New York, NY: HarperCollins.

Milner, B, & Petrides, M. (1984). Behavioral effects of frontal-lobe lesions in man. *Trends in Neuroscience*, 7:403-407.

Minkowski, H. (1952). Space and Time. In H. A. Lorentz, A. Einstein, H. Minkowski, & H. Weyl, *The principle of relativity: A collection of original memoirs on the special and general theory of relativity*. New York, NY: Dover.

Miresco, M. J. & Kirmayer, L. J. (2006). The persistence of mind-brain dualism in psychiatric reasoning about clinical scenarios. *American Journal of Psychiatry,* 163, 913-918.

Mitchell, J. P., Mason, M. F., Macrae, C. N. & Mahzarin, R. B. (2006). Thinking about others: The neural substrates of social cognition. In J. T. Cacioppo, P. S. Visser & C. L. Pickett (Eds.), *Social neuroscience, people thinking about people* (pp. 63-82). Cambridge, MA: MIT Press.

Mitchell, J. P., Macrae, C. N. & Banaji, M. R. (2006). Dissociable medial prefrontal contributions to judgments of similar and dissimilar others. *Neuron*, 50, 655-663.

Morris, J. & Dolan, R. (2004). Emotion and memory: Functional neuroanatomy of human emotion. In R. S. J. Frackowiak, K. J. Friston, C. D. Frith, R. J. Dolan, C. J. Price, S. Zeki, J. Ashburner & W. Penny (Eds.), *Human Brain Function*, 2nd Ed., (pp. 365-396), San Diego, CA: Elsevier Science USA.

Nair, H. P. & Young, L. J. (2006). Vasopressin and pair-bond formation: Genes to brain and behavior. *Physiology*, 21, 145-152.

Neisser, U. (1967). *Cognitive psychology*. New York, NY: Appleton-Century-Crofts.

Niv, Y. (2007). Cost, benefit, tonic, phasic: What do response rates tell us about dopamine and motivation? In B. W. Balleine, K. Doya, J. O'Doherty & M. Sakagami (Eds.). Reward and decision making in corticobasal ganglia networks. *Annals of the New York Academy of Sciences*, 1104, 357-76.

Norris, C. J. & Cacioppo, J. T. (2007). I know how you feel: Social and emotional information processing in the brain. In E. Harmon-Jones & P. Winkielman (Eds.), *Social neuroscience: Integrating biological explanations of social behavior* (pp. 84-105). New York, NY: Guilford Press.

Nowak, M. A. (2006). *Evolutionary dynamics: Exploring the equations of life*. Cambridge, MA: Belknap Harvard University Press.

Ochsner, K. N. (2007). How thinking controls feelings: A cognitive neuroscience approach. In E. Harmon-Jones & P. Winkielman (Eds.), *Social neuroscience: Integrating biological explanations of social behavior* (pp. 106-133). New York, NY: Guilford Press.

Ochsner, K. N. (2006). Characterizing the functional architecture of affect regulation: Emerging answers and outstanding questions. In J. T. Cacioppo, P. S. Visser & C. L. Pickett (Eds.), *Social neuroscience, people thinking about people* (pp. 245-268). Cambridge, MA: MIT Press.

Ochsner, K. N., Bunge, S. A., Gross, J. J. & Gabrieli, J. D. E. (2005). Rethinking feelings: An fMRI study of the cognitive regulation of emotion. In J. T. Cacioppo & G. G. Berntson (Eds.), *Social neuroscience: Key readings in social psychology* (pp. 253-270). New York, NY: Psychology Press.

O'Doherty, J. P., Hampton, A., & Kim, (H. 2007). Model-Based fMRI and its application to reward learning. In B. W. Balleine, K. Doya, J. O'Doherty & M. Sakagami (Eds.). Reward and decision making in corticobasal ganglia networks. *Annals of the New York Academy of Sciences*, 1104, 35-53.

Ogar, J. & Gorno-Tempini, M. L. (2007). The orbitofrontal cortex and the insula. In B. L. Miller & J. L. Cummings (Eds.), *The human frontal lobes*, 2nd Ed. (pp. 59-67). New York, NY: Guilford Press.

Padoa-Schioppa, C. & Assad, J. A. (2006). Neurons in the orbitofrontal cortex encode economic value. *Nature*, 441, 223-226.

Pandya, D. N., & Barnes, C. L. (1987). Architecture and connections of the frontal lobe. In E. Perecman (Ed.), *The frontal lobes revisited* (pp. 41-72) New York, NY: IRBN Press.

Panksepp, J. (1998). *Affective Neuroscience: The foundations of human and animal emotions.* New York, NY: Oxford University Press.

Petrides, M. (2005). Lateral and prefrontal cortex: Architectonic and functional organization. *Philosophical Transactions of the Royal Society*, 360, 781-795.

Petrides, M. & Pandya, D. N. (2002). Association pathways of the prefrontal cortex and functional observations. In D. T. Stuss & R. T. Knight (Eds.), *Principles of frontal lobe function* (pp. 31-50). New York, NY: Oxford University Press.

Phelps, E. A. & LaBar, K. S. (2006). Functional neuroimaging of emotion and social cognition. In R. Cabeza & A. Kingstone (Eds.), *Handbook of functional neuroimaging of cognition* (pp. 421-453). Cambridge, MA: MIT Press.

Phelps, E. A. (2004). The human amygdala and awareness: Interactions between emotion and cognition. In M. S. Gazzaniga (Ed.), *The cognitive neurosciences* III (pp. 1005-1015). Cambridge, MA: MIT Press.

Poldrack, R. A. & Willingham, D. T. (2006). Functional neuroimaging of skill learning. In R. Cabeza & A. Kingstone (Eds.), *Handbook of functional neuroimaging of cognition* (pp. 114-148). Cambridge, MA: MIT Press.

Portas, C., Maquet, P., Rees, G., Blakemore, S. & Frith, C. (2004). Higher cognitive functions: The neural correlates of consciousness. In R. S. J. Frackowiak, K. J. Friston, C. D. Frith, R. J. Dolan, C. J. Price, S. Zeki, J. Ashburner & W. Penny (Eds.), *Human Brain Function* (pp. 269-301) (2nd Ed). San Diego, CA: Elsevier Science USA.

Preuschoff, K. & Bossaerts, P. (2007). Adding prediction risk to the theory of reward learning. In B. W. Balleine, K. Doya, J. O'Doherty & M. Sakagami (Eds.). Reward and decision making in corticobasal ganglia networks. *Annals of the New York Academy of Sciences*, 1104, 135-46.

Raichle, M. E. (2006). Functional neuroimaging: A historical and physiological perspective. In R. Cabeza & A. Kingstone (Eds.), *Handbook of functional neuroimaging of cognition* (pp. 3-20). Cambridge, MA: MIT Press.

Random House: *Webster's College Dictionary* (1997). New York, NK: Random House Inc.

Risberg, J. (2006). Evolutionary aspects on the frontal lobes. In J. Risberg & J. Grafman (Eds.), *The frontal lobes, development, function and pathology* (pp. 1-20). New York, NY: Cambridge University Press.

Roepstorf, A. (2004). Mapping brain mappers: An ethnographic coda. In R. S. J. Frackowiak, K. J. Friston, C. D. Frith, R. J. Dolan, C. J. Price, S. Zeki, J. Ashburner & W. Penny (Eds.), *Human Brain Function* (pp. 1105-17), 2nd Ed. San Diego, CA: Elsevier Science USA.

Rolls, T. R. (2002). The functions of the orbitofrontal cortex. In D. T. Stuss & R. T. Knight (Eds.), *Principles of frontal lobe function* (pp. 354-375). New York, NY: Oxford University Press.

Salloway, S. P. & Blitz, A. (2002). Introduction to functional neural circuitry. In G. B. Kaplan & R. P. Hammer (Eds.) *Brain circuitry and signaling in psychiatry: Basic science and clinical implications* (pp. 1-29). Washington, DC: American Psychiatric Publishing Inc.

Sapir, E. (1949). *Selected writings of Edward Sapir*. Berkeley: University of California Press.

Schlacter, D. (2001). *The Seven Sins of Memory.* New York, NY: Houghton Mifflin.

Schultz, W. & Tremblay, L. (2006). Involvement of primate orbitofrontal neurons in reward, uncertainty, and learning. In D. H. Zald & S. L. Rauch (Eds.), *The orbital frontal cortex* (pp. 173-198). New York, NY: Oxford University Press.

Schmalhofer, F. & Perfetti, C. A. (2007). Neural and behavioral indicators of integration processes across sentence boundaries. In F. Schmalhofer & C. A. Perfetti (Eds.), *Higher level language process in the brain: Inference and comprehension processes* (pp. 161-188). Mahwah, NJ: Lawrence Erlbaum Associates Inc.

Shiffrin, & Atkinson, (1969). Storage and retrieval processes in long-term memory. *Psychological Review,* 76, 179-193.

Shilpa, P., Rodak, K. L., Manikonyan, E., Singh, K. & Platek, S. M. (2007). Introduction to evolutionary cognitive neuroscience methods. In S. M. Platek, J. P. Keenan & T. K. Shackelford (Eds.), *Evolutionary cognitive neuroscience* (pp. 47-62). Cambridge, MA: MIT Press.

Siegel, D. J. (1999). *The developing mind.* New York, NY: Guilford Press.

Skinner, B. F. (1953). *Science and human behavior.* New York, NY: Macmillan Publishing Inc.

Striedter, G. F. (2005). *Principles of brain evolution.* Sunderland, MA: Sinauer Associates

Stuss, D. T., Terence, P. W. & Alexander, M. P. (2001). Consciousness, self-awareness, and the frontal lobes. In P. S. Salloway, P. F. Malloy & J. D. Duffy (Eds.), *The frontal lobes and neuropsychiatric illness* (pp. 101-109). Washington, DC: American Psychiatric Publishing.

Tapiero, I. & Fillon, F. (2007). Hemispheric asymmetry in the processing of negative and positive emotional inferences. In F. Schmalhofer & C. A. Perfetti (Eds.), *Higher level language process in the brain: Inference and comprehension processes* (pp. 355-377). Mahwah, NJ: Lawrence Erlbaum Associates Inc.

Thompson-Schill, S. L., Kan, I. P. & Oliver, R. T. (2006). Functional neuroimaging of semantic memory. In R. Cabeza & A. Kingstone (Eds.), *Handbook of functional neuroimaging of cognition* (pp. 149-190). Cambridge, MA: MIT Press.

Toates, F. M. (2007). *Biological Psychology* (2nd Ed). Harlow, UK: Pearson Education Limited.

Tomasello, M. (2004). What kind of evidence could refute the UG hypothesis? *Studies in Language,* 28, 642-645.

Tomasello, M. (2005). *Constructing a language: A usage-based theory of language acquisition.* Cambridge, MA: Harvard University Press.

Tom, S. M., Fox, C. R., Trepel, C. & Poldrack, R. A. (2007). The neural basis of loss aversion in decision-making under risk. *Science*, 315, 515-518.

Tranel, D. (2000). Electrodermal activity in cognitive neuroscience: Neuroanatomical and neuropsychological correlates. In R. D. Lane & L. Nadel (Eds.), *Cognitive neuroscience of emotion* (p. 218). New York, NY: Oxford University Press.

Tranel, D. (2002). Emotion, decision making, and the ventromedial prefrontal cortex. In D. T. Stuss & R. T. Knight (Eds.), *Principles of frontal lobe function* (pp. 338-353). New York, NY: Oxford University Press.

Tse, D., Langston, R. F., Kakeyama, M., Bethus, I., Spooner, P., Wood, E., Witter, M. & Morris, R. (2007). Schemas and memory consolidation. *Science*, 316, 76-82.

Tversky, A. & Kahneman, D. (1992). Advances in prospect theory: Cumulative representation of uncertainty. *Journal of Risk & Uncertainty*, 5, 297-323.

Vygotsky, L. S. & Luria, A. R. (1993). *Studies on the history of behavior: Ape, primitive, and child.* V. I. Golod & J. E. Knox (Editors & Translators). Mahwah, NJ: Lawrence Erlbaum Associates Inc.

Wagner, A. D., Bunge, S. A. & Badre, D. (2004). Cognitive control, semantic memory, and priming: Contributions from the prefrontal cortex. In M. S. Gazzaniga (Ed.), *The cognitive neurosciences* III (pp. 709-725). Cambridge, MA: MIT Press.

Whorf, B. L. (1956). *Language, thought and reality*. Cambridge, MA: MIT Press.

Wingfield, J. C., More, I. T., Goyman, W., Wacker, D. W. & Sperry, T. (2006). Context and ethology of vertebrate aggression: Implications for the evolution of hormone-behavior interactions. In R. J. Nelson (Ed.), *Biology of Aggression* (pp. 179-210). New York, NY: Oxford University Press.

Wright, J. H. (2004). In J. H. Wright (Ed.), *Cognitive behavioral therapy*. Washington, DC: American Psychiatric Publishing.

Young, L. J. & Wang, Z. (2004). The neurobiology of pair bonding. Nature *Neuroscience*, 7, 1048-1054.

In:Cognitive Science Compendium Volume 2
Editors: Miao-Kun Sun

ISBN: 978-1-60456-775-5

Chapter 5

EXPLORING THE PAST AND IMPENDING FUTURE IN THE HERE AND NOW: MIND-WANDERING IN THE DEFAULT STATE

Malia F. Mason[1*], Moshe Bar[2] and C. Neil Macrae[3]
[1]Columbia Business School, Columbia University, New York, United States of America
[2]Martinos Center for Biological Imaging, MGH, Massachusetts United States of America,
[3]School of Psychology, University of Aberdeen, Aberdeen, United Kingdom

ABSTRACT

Although mind-wandering is a ubiquitous psychological phenomenon, little is known about the neural operations that support this core feature of human cognition. Using a combination of behavioral and functional brain imaging (fMRI) methods, the current investigation demonstrates that recruitment of circumscribed regions of the default network – cortical areas that are active during unconstrained cognitive periods – depends on people's proclivity for reflecting on events from the distant past or impending future while mind-wandering. The present results reveal that whereas medial temporal regions (e.g., parahippocampal cortex) are recruited when the mind wanders to memories of past episodes, areas involved in propsection (e.g., frontal polar and hippocampal) are active when mind-wandering centers on future events. Consistent with recent findings, there is some overlap between the cortical regions involved in thinking about the past and the future. The significance of spontaneous self-projection in time is considered.

Keywords: Mind-wandering, stimulus-independent thought, SIT, Default network, Default state, Prospective memory, medial temporal lobe, Prospection, Simulation, Self-Projection, Mental time travel, goals.

---

1. The other two factors, which accounted for 20% and 22% of the variability in participants' responses, were related to the general proclivity people have for producing and becoming absorbed by these thoughts (i.e.,

Introduction

Although most of us are all too familiar with the mind's proclivity for generating thought that is unbound to the "here and now" (e.g., "considering who to invite to a dinner party while walking to work"), surprisingly little is known about the extensive time periods in which people engage in stimulus-independent thought (SIT). In particular, precisely where does the mind tend to wander to and how does it spontaneously create thoughts that are unrelated to current sensory inputs? Present across cultures (Singer & McCraven, 1962), mind-wandering is hypothesized to constitute a psychological baseline to which people return when they "do nothing," or when they engage in activities that can be accomplished without conscious supervision (Klinger, 1971; Singer, 1966). Recent evidence indicates that a network of cortical regions that is active when people are resting or performing tasks with minimal processing demands – the *default network* (Raichle et al., 2001) – may support this cognitive capacity (Andreasen et al., 1995; Binder et al., 1999; Christoff, Ream, & Gabrieli, 2004; Maguire et al., 2001; Mason et al., 2007; Mazoyer et al., 2001; McGuire et al., 1996). Quite how activity in the default-network relates to mind-wandering content, however, has yet to be elucidated.

Several attempts at specifying where people's minds tend to wander to when they engage in SIT have previously been undertaken. In perhaps the most exhaustive effort to clarify the content of mind-wandering, Klinger and Cox (1987) analyzed more than 1,400 thought samples that were collected from 29 individuals over the course of several days (see also Klinger, 1999; Singer, 1966; Singer & McRaven, 1961). Based on their findings, they concluded that while people occasionally entertain thoughts about improbable events (i.e., they fantasize), the mind more frequently wanders to episodes from the past (e.g., remembering the last time one missed a meeting while brushing one's teeth) and to current practical concerns (e.g., pondering whether one has time to stop for coffee while dressing for work). While under certain circumstances this spontaneous self-projection is problematic, such as when it interferes with current processing goals, there is reason to suspect that it serves an adaptive function in cognition, facilitating problem-solving and planning when circumstances preclude doing something immediate about an existing goal (Bar, 2007; Buckner & Carroll, 2007; Dudai, 2005; Klinger, 1990; Schacter and Addis, 2007; Singer & Antrobus, 1972; Suddendorf & Corballis, 1997; Szpunar, Watson & McDermott, 2007).

As mind-wandering regularly involves these two types of thought (i.e., past vs. impending future), we anticipated that activity in specific regions of the default-network would be associated with these distinct mental contents. Although it would be reasonable to explore this possibility by measuring BOLD changes in the default network while participants respond to prompts that direct them to consider events and people from their past and to reflect on "unfinished business" (cf. Addis, Wong & Schacter, 2007), one would possibly be doing so at the expense of the hallmark features of this form of mentation: its relatively unbidden, undirected and unmonitored (e.g., "Am I still thinking about events from the past?") nature. In light of evidence that people have a "mind-wandering style" (cf. Singer & Antrobus, 1963; 1972) – a favored pattern of thought that they default to when the external environment ceases to pose challenging and variable processing demands – the approach taken in the present investigation was to develop a model of mind-wandering content by factor-analyzing participants' responses to the Imaginal Process Inventory (IPI; Singer & Antrobus, 1972), a questionnaire designed to measure aspects of inner thought, and then relate participants' mind-wandering content factor scores to the default network activity they exhibit during periods of frequent mind-wandering.

Given the high level of resting-state metabolism observed in the same medial temporal lobe regions that support episodic memory (Greicius, Kransow, Reiss & Menon, 2002; Greicius, Srivastava, Reiss & Menon, 2004), we expected that activity in this area during high incidence SIT periods would be greatest among individuals whose mind-wandering episodes tend to dwell on episodes and people of the past. In contrast, we expected that people with a proclivity for reflecting on impending events or "current concerns" (Klinger, 1971) while mind-wandering would exhibit significantly greater recruitment of both frontal regions that are implicated in such things as reflecting on intentions to act in the future (e.g., BA10; Okuda et al., 2003) and medial temporal regions implicated in retrieving details of past episodes. This latter hypothesis stems, in part, from recent evidence indicating that simulating possible future scenarios requires retrieving contextual details from past happenings (see Addis, Wong & Schacter, 2007; Bar, Aminoff, Mason & Fenske, 2007; Hassavis, Kumaran, Vann & Maguire, 2007; Okuda et al., 2003; Szpunar, Watson & McDermott, 2007).

Materials and Methods

The approach taken in the current investigation was to obtain measures of impending future and past- oriented mind-wandering and then to relate these constructs to the activity observed in the default network when people engage in SIT. Prior to conducting the fMRI portion of the experiment, we administered a questionnaire that measures mind-wandering and then used factor analysis to identify dimensions related to mind-wandering content to regress against our functional imaging data. Identifying cortical regions associated with impending future and past-oriented mind-wandering involved a series of 3 experimental phases (see Mason et al., 2007 for a detailed description). In Phase 1, we established high incidence SIT periods by extensively training participants on a small set of working-memory sequences and then confirmed that people experience a greater incidence of SIT while performing the trained sequences relative to new sequences (Phase 2). In Phase 3, we identified cortical regions (with fMRI) that were more active during high relative to low incidence SIT periods (when participants performed 'trained' sequences relative to 'new' sequences) and then regressed participants' mind-wandering content scores against this pattern of cortical activity. To confirm that people have relatively stable mind-wandering styles, we administered the IPI on a second occasion, approximately 32 months later, and assessed the reliability of their responses over a period of several years.

Obtaining Measures of Impending Future and Past-Oriented Mind-Wandering

Participants

A total of 194 individuals (127 female) participated in this portion of the experiment in exchange for course credit. All participants gave informed consent according to the procedures approved by the Committee for the Protection of Human Subjects (CPHS).

Design and Stimulus Materials

To confirm that the mind tends to wander to events from the past and to immediate future concerns, we administered portions of the IPI (Singer & Antrobus, 1972) and conducted a factor analysis on participants' responses. Participants were administered a total of 140 attitudinal statements (e.g., "*I sometimes daydream about people and places I was familiar with when young*") from 11 subscales related to daydreaming to which they were asked to indicate the extent to which they agreed or disagreed (1: *strongly uncharacteristic* to 5: *strongly characteristic*).

Procedure

Each participant completed an electronic version of the questionnaire. It was explained that the questionnaire pertained to mind-wandering, or their spontaneous thoughts and daydreams. Participants were told that, within the context of the present experiment, the term 'daydream' referred to the "shifting of attention away from an ongoing physical or mental task or from a perceptual response to some internal stimulus" and provided an example: "*thinking about an upcoming party while making a tuna sandwich.*"

Factor analysis was conducted using principle component analysis followed by a varimax rotation (Goldberg & Digman, 1994). As we anticipated finding high correlations among the constructs measured by the IPI subscales (e.g., 'Present Oriented Thought' and 'Problem-Solving' in Daydreams) and because we were interested in how these higher-order factors related to cortical activity in the default network, the analyses were undertaken on the IPI subscale means. Results of Cronbach's reliability analyses confirmed that each of the eleven subscales was comprised of items related to a single uni-dimensional construct. Eight of the eleven subcales on the IPI had high internal consistency (Cronbach's Alpha > .80). The remaining three had adeqsuate alpha levels of .73, .74, and .72. Components were restricted to those with Eigenvalues > 1.

Results

Results indicated that 67% of the variance in participants' responses to the questionnaire was accounted for by four factors. Of particular interest to the present investigation were the two factors related exclusively to mind-wandering content (see Table 1 for factor loadings and example items). The first of these factors – *propensity for impending future thought* – accounted for 14% of the variance in participants' responses, with high positive loadings on scales measuring the tendency to reflect on immediate concerns and generate potential solutions to unresolved problems and negative loadings on the scale measuring improbable thinking while daydreaming. The second mind-wandering content factor – *propensity for past mind-wandering* – had high loading on the scale measuring the tendency to reflect on the past and accounted for 11% of the variance.[1] Having established these two factors, we then computed factor scores for participants in the fMRI portion of the experiment using the rotated loadings as weights.

Table 1.0 Varimax rotated component matrix displaying loadings greater than 0.5 and less than -0.5 for the factors related to mind-wandering content

Subscale	Sample Items	Impending Future	Past
'Present Orientation in Daydreams'	"My present day concerns are usually reflected in my daydreams"	.71	
'Problem Solving in Daydreams'	"Thoughts I have are often about different ways of finishing things I still have to do in my life"	.80	
'Bizzare/Improbable Daydreams'	"I often have thoughts about things that could rarely occur in real life"	-.57	
'Past Orientation in Daydreams'	"I sometimes think about people and places I was familiar with when I was young"		.87
	Variance Explained	*14%*	*11%*

ASSESSING THE STABILITY OF PEOPLE'S IMPENDING FUTURE AND PAST-ORIENTED MIND-WANDERING

To confirm that people have relatively stable mind-wandering styles, we administered the IPI on a second occasion, approximately 32 months later, and assessed the reliability of participants' responses over a period of several years.

Participants

Each of the 194 individuals who completed the IPI were contacted 32 months after the initial administration date and asked to complete the questionnaire in exchange for monetary compensation. Approximately 35% of the individuals (N = 69) completed and returned the electronic version of the questionnaire to the experimenter.

Design and Stimulus Materials

To assess the stability of people's mind-wandering experiences (i.e., to establish that people have a mind-wandering "style"), we computed participants' *propensity for impending future thought* and *propensity for past mind-wandering* factor scores. This was done by taking the standardized mean score for each scale, multiplying it by the corresponding factor loading of the variable for the given factor (from the factor model generated 32 months previously) and summing these products.

Results

As expected, results revealed a strong positive correlation between participants *propensity for impending future thought* factor scores computed after the initial administration (Time 1) and their *propensity for impending future thought* factor scores 32 months later (Time 2), $r(68) = .66$, $p < .0001$. Correlations between mean responses to the particular scales with high loadings ($x > .5$ and $x < -.5$) on the *propensity for impending future thought* factor scores at Time 1 and Time 2 were also computed. Results revealed a strong positive correlation between Time 1 and Time 2 for all three scales: 'Problem Solving in Daydreams', $r(68) = .51$, $p < .0001$; 'Present-Oriented Daydreams', $r(68) = .60$, $p < .0001$; and 'Bizarre and Improbable Daydreams' (which was negatively correlated with the factor), $r(68) = .71$, $p < .0001$. Similarly, results revealed a strong positive correlation between participants' Time 1 *propensity for past mind-wandering* and their Time 2 *propensity for past mind-wandering* factor scores, $r(68) = .60$, $p < .0001$. The correlation between mean responses to the one scale with high loadings on the *propensity for past mind-wandering* factor scores – 'Past-Oriented Daydreams' -- at Time 1 and Time 2 was also significant, $r(68) = .65$. $p < .0001$.

Having established that mind-wandering styles can be reliably characterized by the proclivity for considering past and impending future events, we then sought to relate these two dimensions to the activity observed in the default network when people engage in SIT. Identifying cortical regions associated with impending future and past-oriented mind-wandering involved a series of 3 experimental phases (see Mason et al., 2007 for a detailed description). In Phase 1, we established high incidence SIT periods by extensively training participants on a small set of working-memory sequences and then confirmed that people experience a greater incidence of SIT while performing the trained sequences relative to new sequences (Phase 2). In Phase 3, we identified cortical regions (with fMRI) that were more active during high relative to low incidence SIT periods (when participants performed 'trained' sequences relative to 'new' sequences) and then regressed participants mind-wandering content scores against this pattern of cortical activity.

PHASE 1: ESTABLISHING HIGH INCIDENCE MIND-WANDERING PERIODS

In Phase 1, we sought to establish high incidence SIT periods by extensively training participants on a small set of working-memory sequences (cf. Teasdale et al., 1995).

Participants

Nineteen individuals (12 female) who filled out the IPI also completed the training, thought-sampling and fMRI phases of the experiment in exchange for course credit. All participants were right-handed, native English speakers with no history of neurological problems. Participants gave informed consent according to the procedures approved by the CPHS.

Procedure

Training (Days 1-3)

During Phase 1, participants were trained on both a verbal and a motor working-memory task. The verbal working-memory task involved remembering and manipulating 4 four-letter sequences (e.g., 'R H V X'). A verbal working memory trial consisted of the following sequence of events. The target sequence appeared on the screen for a duration of 800ms, at which point it was replaced by an arrow which indicated the direction participants should reference the sequence (an arrow to the 'left' indicated the participant should think of the sequence in reverse – as 'X V H R' – while an arrow to the 'right' indicated that the string should be references in the forwards direction). Participants were then shown each of the letters contained in the string, to which they indicated, via a key-press, the position of the letter (e.g., if they saw 'R H V X' after a 'left' arrow the correct response would be '2'). The motor task involved remembering and manipulating 4 finger-tapping patterns. A motor trial consisted of the following sequence of events. Four boxes appeared at the center of the screen and illuminated, one at a time, in the target key sequence order. Eight hundred milliseconds later, an arrow indicating the direction in which they should repeat the target sequence appeared on the screen. An arrow to the right indicated that participants should repeat the sequence in the order just displayed, while an arrow to the left signaled that the key sequence should be repeated in the reverse order. All trials in both the verbal and motor working-memory task were self-paced, leaving no time in any of the block for participants to "do nothing". On three consecutive days, participants reported to the laboratory for 30 minutes of training with the 4 letter and 4 finger-tapping sequences.

Phase 2: Thought-sampling (Day 4)

The goal of Phase 2 was to confirm that people experience a greater incidence of SIT while performing the trained sequences relative to new sequences. Upon arrival on the fourth day of training, the 19 participants were informed that they again would be practicing the learned sequences in blocks but that, at random intervals, they would be interrupted and asked to indicate, via a binary key-press, whether they were experiencing SIT (defined for them as "thoughts that do not facilitate performance and are not immediate reactions to perceptual information gleaned over the course of a trial"). Participants were then informed of one additional procedural change. In addition to performing blocks of the working-memory task on the eight learned sequences (4 letter strings, 4 motor patterns), they would also be performing the task on sequences that they had not practiced previously. Each participant was then administered three 7-min runs of interleaved rest (i.e.,'+') and task blocks (i.e., verbal and motor working memory tasks). Data from one participant were not included in the analysis because of a computer malfunction.

Results

Consistent with results of other studies (e.g., Teasdale et al., 1995), the propensity to mind wander was most pronounced when participants performed practiced sequences (i.e., when the task could be performed without conscious supervision) and during rest periods relative to periods in which novel sequences were performed [$F(2,34) = 81.49$, $p < .01$]. Participants reported a greater incidence of mind wandering during the baseline (i.e., rest) blocks ($M = .93$; $SD = .16$) than during both practiced blocks ($M = .32$, $SD = .20$), $t(17) = 9.22$, $p < .01$, and novel blocks ($M = .22$, $SD = .18$), $t(17) = 10.96$, $p < .01$. More importantly, however, participants reported a significantly greater incidence of mind wandering during the practiced blocks than during the novel blocks [$t(17) = 2.11$, $p < .05$], even though the tasks were identical.

Phase 3: Identifying Neural Correlates of Past and Impending Future Mind-Wandering

In Phase 3, we identified cortical regions (with fMRI) that were more active during high relative to low incidence SIT periods (when participants performed ‘trained’ sequences relative to ‘new’ sequences) and then regressed participants mind-wandering content scores against this pattern of cortical activity.

Procedure

Functional MRI (Day 5)

On the fifth and final day of the experiment, participants were scanned (fMRI) while performing both working memory tasks on the practiced sequences, on novel sequences and while passively observing a centrally-presented fixation cross. The experiment consisted of 5 EPI scans, each lasting 6 minutes and 50 seconds. Each of the blocks (fixation, practiced sequences, novel sequences) within a single scan lasted between 20 and 40 seconds in duration.

Images were acquired using a 1.5 Tesla whole body scanner (GE Signa) with a standard head coil. Visual stimuli were generated with Presentation software (www.neuro-bs.com) on a Dell computer. Stimuli were projected to participants with an Epson (model ELP-7000) LCD projector onto a screen positioned at the head end of the bore. Participants viewed the screen through a mirror. Cushions were used to minimize head movement.

T1- weighted anatomical images were collected using a high-resolution 3-D sequence (SPGR; 128 sagital slices, TR = 7 ms, TE = 3 ms, prep time = 315 ms, flip angle = 15°, FOV = 24 cm, slice thickness = 1.2 mm, matrix = 256 x 192). Functional images were collected in runs using a gradient echo EPI sequence (each volume comprised 25 slices; 4.5 mm thick, 1 mm skip; TR = 2500 ms, TE = 35 ms, FOV = 24 cm, 64 x 64 matrix; 90° flip angle).

Image Analysis

Functional MRI data were analyzed using SPM99 (Friston et al., 1995). For each functional run, data were preprocessed to remove sources of noise and artifact. Functional data were corrected for differences in acquisition time between slices for each whole-brain volume, realigned within and across runs to correct for head movement, and co-registered with each participant's anatomical data. Functional data were then transformed into a standard anatomical space (3 mm isotropic voxels) based on the ICBM 152 brain template (MNI) which approximates Talairach and Tournoux's atlas space. Normalized data were then spatially smoothed using a Gaussian kernel (6 mm FWHM).

For each participant, a general linear model specifying task effects (modeled with a function for the hemodynamic response), runs (modeled as constants), and scanner drift (modeled with linear trends) was used to compute parameter estimates (β) and *t*-contrast images for each comparison at each voxel. These individual contrast images were then submitted to a second-level, random-effects analysis to obtain mean *t*-images (thresholded at $p = .001$, uncorrected; $k = 10\ mm^3$). An automated peak-search algorithm identified the location of peak activations based on *z*-value and cluster size.

Results

Neural Correlates of Mind-wandering

The default-network was functionally defined by comparing the neural response associated with rest (i.e., baseline) to the response associated with all-task periods (i.e., novel and practiced blocks). This comparison revealed significantly greater recruitment in the: posterior cingulate and precuneus, posterior lateral cortices, bilateral insula, anterior cingulate, bilateral parahippocampal gyri, and aspects of ventral and dorsal MPFC (Mazoyer et al., 2001; Shulman et al., 1997). To identify default-network regions that were more active during periods of frequent relative to infrequent mind wandering, we compared neural activity when participants performed practiced blocks to activity during novel blocks using the 'baseline > task' contrast as an inclusive mask. As predicted, activity in nearly all default-network regions was significantly greater during frequent mind-wandering periods (Figure 1). Importantly, no default-network regions exhibited greater activity during periods of infrequent mind-wandering (i.e., 'novel > practiced', inclusively masked with 'baseline > task').

Neural Correlates of Impending Future and Past-Oriented Mind-wandering

To assess the relationship between default-network activity and mind-wandering content, we computed participants' factor scores on the experimental dimensions of interest (i.e., *propensity for impending future thought* & *propensity for past thinking*). The resulting scores were then regressed against the pattern of activity participants exhibited during high incidence mind-wandering periods (i.e., 'practiced > novel', inclusively masked with 'baseline > task'; $p < .05$, k=15; see Table 2). Consistent with our predictions, the greater the self-reported tendency to mind-wander about impending events and unresolved matters, the greater the activity observed in a cluster that extended across bilateral medial frontal (BAs 10,8) and superior frontal (BA9) gyri, a cluster that extended across the left hippocampus and amygdala (BAs 24,28,36), and a cluster in the left parrahippocampal (BA36)/fusiform gyri (BA37). There were also significant positive correlations detected in the R. anterior cingulate (BAs 24,31), in the R. supramarginal gyrus (BA40) and the L. claustrum (BA37). No cortical regions exhibited a negative correlation with this factor. Furthermore, the stronger the tendency to mind wander about events from the past, the greater the activity observed in the R. parahippocampus (BA35,36), R. insula (BA13), and the R. cingulate (BA24). Significant negative correlations with *propensity for past thinking* were observed in bilateral medial frontal regions (BAs 9 and10) and in the posterior cingulate (BA31).

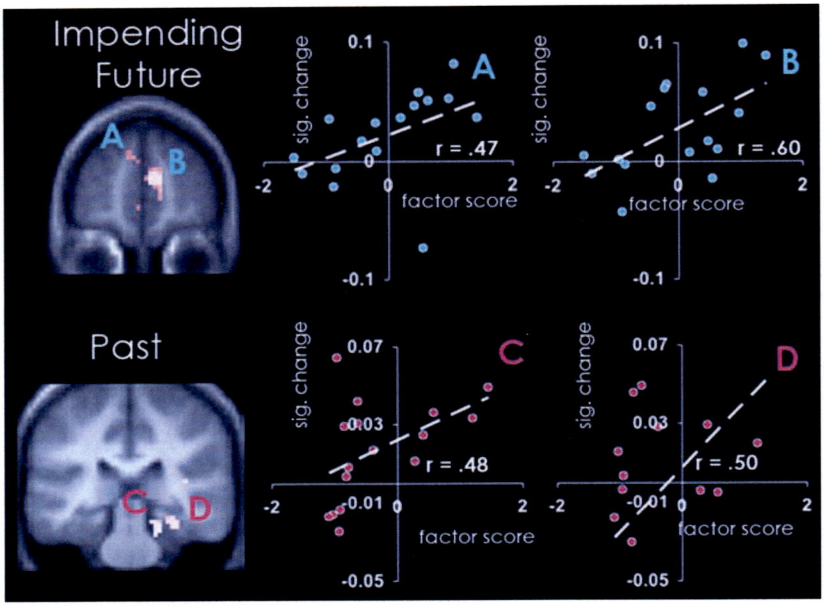

Figure 1. Default-network regions associated with thinking about the impending future (i.e., "concern-related" thought) and the past while mind-wandering. Depicted are regions within the default-network [i.e., areas that emerged from the contrast 'rest > task' (p<.05, k=10)] that exhibited greater activity during frequent relative to infrequent mind-wandering periods (p<.001, k=10) and that correlated positively with participants' *propensity for impending future thought* scores (top) and *propensity for past mind-wandering* scores (bottom). Scatter-plots depict participants' BOLD difference scores (practiced – novel) plotted against their factor scores. BOLD signal values for the two blocks were computed for each participant by averaging the signal in regions within 8 mm of the peak from 3 TRs (7.5 s) until 7 TRs (17.5 s) after the block onset. 'A'= R. mPFC (BA9; -9,54,28; $p = .003$, uncorrected); 'B' = X (BA10; 6,53,13; $p = .003$, uncorrected); 'C' = R. parahippocampus (BA36; 21,-30,-19 $p = .006$, uncorrected); and 'D'= R. parahippocampus (BA36; 30,-33,-14; $p = .006$, uncorrected).

DISCUSSION

Previous functional imaging and lesion studies suggest that the cued retrieval of episodes from the past and future involve overlapping and unique aspects of medial temporal lobe and frontal polar regions (Addis et al., 2007; Okuda et al., 2003; 2005; Szpunar, Watson & McDermott, 2007). Of particular interest is recent evidence revealing that imagining future events involves frontal polar areas that have been implicated in prospective memory tasks (e.g., Burgess, Quinn & Frith, 2001; Okuda et al., 2003) as well as areas, like the hippocampus, which are thought to support episodic memory (e.g., Hassabis, Kumaran, Vann & Maguire et al., 2007). This pattern of findings is interpreted as evidence that mentally envisioning the future involves retrieving memories of similar past episodes or contextually relevant information (Atance & O'Neill, 2001; 2005; Bar, Aminoff, Mason & Fenske, 2007; Tulving, 2002). Extending these findings, the current results show that the neural activity observed in these areas when people's thoughts are unconstrained (i.e., during high incidence mind-wandering periods) depends on their proclivity for thinking about the past and reflecting on current concerns.

Table 2. Regions of the default network that exhibited significant positive correlations with 'proclivity for past mind-wandering' and 'proclivity for future mind-wandering' factor scores. Coordinates are reported in Talairach space. The displayed *t*-values are associated with the area's peak hemodynamic response. All coordinates emerged with at a threshold of $p < .05$, k = 15.

			coordinates		
Anatomical Location	**BA**	**k**	**x**	**y**	**z**
Positive Correlations					
L. hippocampus	24/28	17	-20	-13	-17
L. fusiform/parahippocampal gyrus	37/36	29	-30	-43	16
B. medial frontal	10	114	6	54	14
R. supramarginal gyrus	40	30	48	-62	36
B. cingulate gyrus	31	142	6	-15	45
L. clausturm		65	-51	-3	-9
Negative Correlations					
None.					

			coordinates		
Anatomical Location	**BA**	**k**	**x**	**y**	**z**
Positive Correlations					
R. parahippicampal cortex	35/36	42	21	-30	-19
R. insula		21	36	-26	12
B. cingulate gyrus	24	22	9	-6	47
Negative Correlations					
B. medial frontal	9/10	127	-9	68	16
L. posterior cingulate	31	20	-9	-42	35

Although, at first glance, the mind's proclivity for wandering may be worrisome (Schooler, Reichle & Halpern, 2004; Smallwod & Schooler, 2006), it is important to note that the capacity to relive past events and imagine future scenarios may, on the contrary, be quite advantageous (see Buckner & Carrol, 2007; Mason et al., 2007). Through "mental time travel" (Tulving, 2001) – traveling backwards and forwards in subjective time – people can anticipate future events, reconsider past episodes, and formulate strategies and plans based on previous experiences (Bar, 2007; Buckner & Carroll, 2007; Gilbert & Wilson, 2007; Mason et al., 2007; Shallice & Burgess, 1991). This capacity enables humans to consider potential consequences prior to acting (i.e., to simulate), to maintain intentions to act in particular ways over long time periods, and to override momentary needs in pursuit of longer-term goals and objectives (Suddendorf & Corballis, 1997). From this perspective, allocating one's attention away from the present sensory environment to people, places, and events of the past or future (i.e., mind-wandering) is arguably the most adaptive way to use surplus processing resources.

One limitation of the approach taken in the present investigation – regressing participants' factor scores against the BOLD activity they exhibit during frequent SIT – is that it may lack the sensitivity to detect significant effects in regions that subserve future and past-oriented mind-wandering. Indeed, it is quite possible that the present analysis fails to implicate all of the regions involved in producing impending future and past-oriented thought because of a modest sample size, an inadequate range of factor scores or because of an imperfect mapping between the constructs represented by the factors and the functions subserved by the critical cortical areas. Although we believe the present approach provides insight into the process by which people travel back-and-forth in subjective time, by no means do we think it provides an exhaustive account of these mechanisms. For example, it is worth noting that some of the regions that appear to play a general role in introspectively-oriented mental activity (e.g., MPFC; Mason et al., 2007) did not correlate positively with *both* the experiential dimensions of interest (i.e., *propensity for impending future thought* & *propensity for past thinking*). Furthermore, the present investigation did not detect a relationship between participants' factor scores along the dimensions of interest and a couple of areas that would be expected to be involved, such as the *propensity for past thinking* and the hippocampus (cf. Andreasen et al., 1995; Maguire et al., 2001). We suspect that a more complete account of the process by which people divert their attention from the external sensory environment and consider events from their past or "unfinished business" will emerge only through a convergence of approaches.

In considering why the mind evolved the capacity to spontaneously fluctuate among the immediate sensory environment, events of the near and distant past, and possible future scenarios it is helpful to imagine what life would be like without this capacity. One need only reflect on the extensive time periods spent confined to endless business meetings, "on hold" with telephone customer service representatives, and at a stand-still in rush-hour traffic to appreciate the freedom this type of mentation affords. For better or worse, lacking this capacity, we would each be trapped in the present moment. Despite the popularity of "mental time travel" as a strategy for coping with invariable and impoverished settings, it is difficult to imagine that this type of thought *evolved* to serve a coping function. Instead, it seems more plausible that the adaptiveness of this capacity centers on the benefits conferred by engaging in simulation (for a discussion, see Buckner & Carroll, 2007). Whereas simulating past experiences reinforces learning, clarifies uncertainties and deepens one's understanding of previous episodes; through mentally exploring the future one can consider alternatives before acting and maintain intentions to act over quite extensive time periods (e.g., "I have to move my car on the third Thursday of the month for street cleaning"). While the relationship between these two forms of self-projection needs to be clarified, the results of the present study support the idea that, at any given moment in time, people allocate significant processing resources not just to their immediate surroundings but to episodes long past and yet to occur.

ACKNOWLEDGMENTS

The authors would like to thank John D. Van Horn for providing advice and analysis assistance and Jasmine Boshyan and Rebecca Dyer for their help with data collection. This research was supported by a Faculty Development grant from Columbia Business School and Columbia University (to MFM), the National Institute of Neurological Disorders and Stroke Grants NS050615 (to MB), and by a Royal Society-Wolfson Fellowship (to CNM).

REFERENCES

Addis, D. R., Wong, A. T. & Schacter, D. L. (2007). Remebering the past and imaging the future: Common and distinct neural substrates during event construction and elaboration. *Neuropsychologia, 45,* 1363-1377.

Andreasen, N.C., O'Leary, D.S., Cizadlo, T., Arndt, S., & Rezai, K. (1995). Remembering the past: Two facets of episodic memory explored with positron emission tomography. *American Journal of Psychiatry, 152*, 1576–1585.

Bar, M. (2007). Using analogies and associations to generate predictions. *Trends in Cognitive Sciences, 11,* 280-289.

Bar, M., Aminoff, E., Mason, M. & Fenske, M. (2007). The units of thought. *Hippocampus, 17*, 420-428.

Binder, J. R., Frost, J. A., Hammeke, T. A., Bellgowan, P. S., Rao, S. M., & Cox, R. W. (1999). Conceptual processing during the conscious resting state. A functional MRI study. *Journal of Cognitive Neuroscience, 11*, 80-95.

Buckner, R.L. & Carroll, D.C. (2007). Self-projection and the brain. *Trends in Cognitive Science, 11*, 49-57.

Christoff, K., Ream, J.M., & Gabrieli, J.D.E. (2004) Cognitive and neural basis of spontaneous thought processes. *Cortex*, 40, 623-630.

Friston, K. J., Holmes, A. P., Worsley, K. J., Poline, J-P., Frith, C. D., & Frakowiak, R. S. J. (1995). Statistical parametric maps in functional imaging: a general linear approach. *Human Brain Mapping, 2*, 189-210.

Gilbert, D. M. & Wilson, T. D. (2007). Prospection: Experiencing the Future. *Science, 317,* 1351-1354.

Goldberg, L. R. & Digman, J. M. (1994). Revealing structure in the data: Principles of exploratory factor analysis. In S. Strack & M. Lorr (Eds.), Differentiating normal and abnormal personality (pp. 216-242). New York: Springer.

Greicius, M.D., Srivastava, G., Reiss, A.L., & Menon, V. (2004).Default-mode network activity distinguishes Alzheimer's disease from healthy aging: Evidence from functional MRI. *Proceedings of the National Academy of Sciences,* 101, 4637-4642.

Greicius, M.D., Krasnow, B., Reiss, A.L., & Menon, V. (2003). Functional connectivity in the resting brain: a network analysis of the default mode hypothesis. *Proceedings of the National Academy of Sciences, 100,* 253-258.

Klinger, E. (1971). Structure and Functions of Fantasy. New York: Wiley.

Klinger, E. (1990). Daydreaming: Using waking fantasy and imagery for self-knowledge and creativity. New York: Saint Martin's Press.

Klinger, E. (1999). Thought flow: Properties and mechanisms underlying shifts in content. In J. Singer & P. Salovey (Eds.), At play in the fields of consciousness: Essays in honor of Jerome Singer (pp. 29–50). Mahwah, NJ: Erlbaum.

Klinger, E. & Cox, W.M. (1987). Dimensions of thought flow in everyday life. *Imagination, Cognition and Personality, 7*, 105- 128.

Maguire, E. A. (2001). Neuroimaging studies of autobiographical event memory. *Philosophical Transactions of the Royal Society B: Biological Sciences, 356*, 1441-1451.

Mason, M.F., Norton, M.I., Van Horn, J., Wegner, D.M., Grafton, S. F., & Macrae, C.N. (2007). Wandering minds: The default-network and stimulus-independent thought. *Science, 315,* 393-395.

Mazoyer, B., Zago, L., Mellet, E., Bricogne, S., Etard, O., Houde, O., Crivello, F., Joliot, M., Petit, L., & Tzourio-Mazoyer, N. (2001). Cortical networks for working memory and executive functions sustain the conscious resting state in man. *Brain Research Bulletin, 54*, 287-298.

McGuire, P.K., Paulesu, E., Frackowiak, R. S., & Frith, C.D. (1996). Brain activity during stimulus independent thought. *NeuroReport, 7,* 2095-2099.

Okuda, J., Fujii, T., Ohtake, H., Tsukiura, T., Tanji, K., Suzuki, K. et al. (2003). Thinking of the future and the past: The roles of the frontal pole and the medial temporal lobes. *Neuroimage, 19,* 1369-1380.

Schooler, J.W., Reichle, E.D., Halpern, D.V. (2004). Zoning out while reading: Evidence for dissociations between experience and metaconsciousness. In D.T. Levin (Ed) Thinking and Seeing: Visual Metacognition in Adults and Children. (pp. 203-226) Cambridge, MA: MIT Press.

Shulman, G.L., Fiez, J.A., Corbetta, M., Buckner, R. L., Miezin, F.M., Raichle, M.E., & Petersen, S.E. (1997). Common blood flow changes across visual tasks: II. Decreases in cerebral cortex. *Journal of Cognitive Neuroscience, 9*, 648-663.

Shallice, T. & Burgess, P.W. (1991). Deficits in strategy application following frontal lobe damage in man. *Brain, 114,* 727-741.

Singer, J. L. (1966) Daydreaming. New-York: Random House.

Singer, J. L. & Antrobus, J. S. (1963). A factor analytic study of daydreaming and conceptually-related cognitive and personality variables. *Perceptual and Motor Skills,* (Monograph Supplement 3-V17).

Singer, J. L. & Antrobus, J. S. (1972). Daydreaming, imaginal processes, and personality: A normative study. In P. W. Sheehan (Ed.), The nature and function of imagery (pp. 175-202). New York: Academic Press.

Singer, J.L. & McRaven, V. (1961). Some characteristics of adult daydreaming. *Journal of Psychology, 51,* 151-164.

Singer, J.L. & McRaven, V. (1962). Patterns of daydreaming in American subcultures. *International Journal of Social Psychiatry, 8*, 272-282.

Smallwood, J.M. & Schooler, J.W. (2006). The restless mind. Psychological Bulletin, 132, 946-958.

Suddendorf, T. & Busby, J. (2003). Mental time travel in animals. *Trends in Cognitive Sciences, 7,* 391-369.

Sucdendorf, T. & Corballis. M (1997). Mental time travel and the evolution of the human mind. *Genetic, Social, and General Psychology Monographs, 123, 133*-167.

Teasdale, J.D., Dritschel, B.H., Taylor, M.J., & Proctor, L. (1995). Stimulus-independent thought depends on central executive resources. *Memory and Cognition, 23,* 551-559.

Tulving, E. (1985). Memory and Consciousness. *Canadian Psychology, 25,* 1-12.

In:Cognitive Science Compendium Volume 2
Editors: Miao-Kun Sun

ISBN: 978-1-60456-775-5

Chapter 6

Control of Cortical Synaptic Receptive Fields by Neuromodulation and Patterned Sensory Input

Naveen Nagarajan, Michael M. Merzenich, Christoph E. Schreiner and Robert C. Froemke*

Coleman Memorial Laboratory and W.M. Keck Foundation Center for Integrative Neuroscience, Department of Otolaryngology
University of California, San Francisco, California 94143, USA

Abstract

Cortical receptive fields are dynamic representations of sensory information. Studies in animals and humans have determined that the organization of receptive fields is continuously modified by experience and behavior throughout life. The mechanistic principles and synaptic modification rules underlying changes to receptive fields are beginning to be understood. In particular, new research has begun to reveal how neuromodulators and sensory inputs govern the dynamics of synaptic receptive field plasticity. Here we review the field of cortical receptive field plasticity, focusing on new findings from intracellular recordings in vivo, and discuss the broad implications on cognitive functions and neurodegenerative disorders.

Introduction

Sensory information is dynamically processed in the cerebral cortex. In primary sensory cortex, this information is represented by receptive fields at the level of individual neurons, while across the population, neuronal ensembles are organized into stimulus feature maps

* Corresponding author: Dr. R.C. Froemke, Phone: 415-476-1762, Fax: 415-476-1941, Email: rfroemke@phy.ucsf.edu

(Schreiner & Winer, 2007). These maps are spatial structures with a topography reflecting that of the sensory epithelia; e.g., primary auditory cortex (A1) is cochleotopic, while primary visual cortex (V1) is retinotopic. Neuronal receptive fields and feature maps in the cortex are constructed and undergo elaborate modification with neural development (Katz & Shatz, 1996) and novel experience over the lifespan of an animal (Buonomano & Merzenich, 1998). While subcortical sensory brain regions also have highly ordered neuronal receptive fields and can contain sensory maps (Winer & Schreiner, 2005), the cortex has been the subject of numerous studies in part because cortical circuits are quite plastic: cortical neurons and networks possess a high capacity to change in response to alterations of sensory experience or patterns of electrical activity (Hensch, 2005; Karmarkar & Dan, 2006; Weinberger, 2004).

Measurements of cortical receptive fields and maps have been conducted for different sensory systems in a wide variety of mammalian species, including mice, rats, primates, and humans. These empirical studies have shown that cortical networks undergo rapid reorganization in response to either a novel behavioral experience or to highly structured input from the periphery. Individual synapses within cortical neuronal networks contribute to and control receptive field plasticity and map reorganization, as modification of synaptic input directly determines changes in spiking output. These experience-dependent modifications at the synaptic level include increases, decreases, and maintenance of synaptic strength (Malenka & Bear, 2004). Persistent enhancement of synaptic strength is referred to as long-term potentiation (LTP), while persistent suppression is called long-term depression (LTD). In most cases, long-term modification of synaptic weights and receptive fields requires the activation of subcortical neuromodulatory systems, which normally convey the behavioral meaning or relevance of certain sensory stimuli (Froemke et al., 2007; Gu, 2002; Weinberger, 2004). Decades of research have attempted to decipher the relationship between modifications to individual synapses and cortical map changes that occur at the network level (Buonomano & Merzenich, 1998). Understanding the mechanistic link between neural dynamics and self-organization of cortical maps requires a clear description of three factors: 1) the neural computations performed by cortical circuits, 2) the synaptic mechanisms of plasticity, and 3) the interactions between these elements that lead to behavioral modification. While most previous studies relied on extracellular electrophysiological recordings of spike output, recent advances in understanding cortical organization and plasticity have been achieved with intracellular recording techniques. In this review, we focus on new results connecting synaptic and receptive field plasticity in A1 and V1, as models for relating activity at the synaptic and systems levels, and with the aim of understanding how low-level processes such as synaptic transmission and plasticity support higher-level cognitive functions including perception, arousal, attention, and learning and memory, and might be linked to nervous system disorders.

CORTICAL RECEPTIVE FIELD ORGANIZATION

Cortical receptive fields are the neuronal representations of sensory information. Individual neurons are tuned to specific stimulus features of the sensory environment (Figure 1); for example, in A1, neurons are tuned to certain acoustic properties such as tone frequency or pitch (Imaizumi et al., 2004), intensity (Tan et al., 2007), direction of change (Zhang et al.,

2003), and perhaps some forms of vocalization (Liu & Schreiner, 2007). Neuronal tuning curves can vary in terms of best stimulus identity and tuning width, but in general, the structure of tuning curves is somewhat stereotyped, consisting of a single peak, defining the best feature, and having a semi-Gaussian or inverted 'V' shape, defining the width and degree of feature selectivity for each cell.

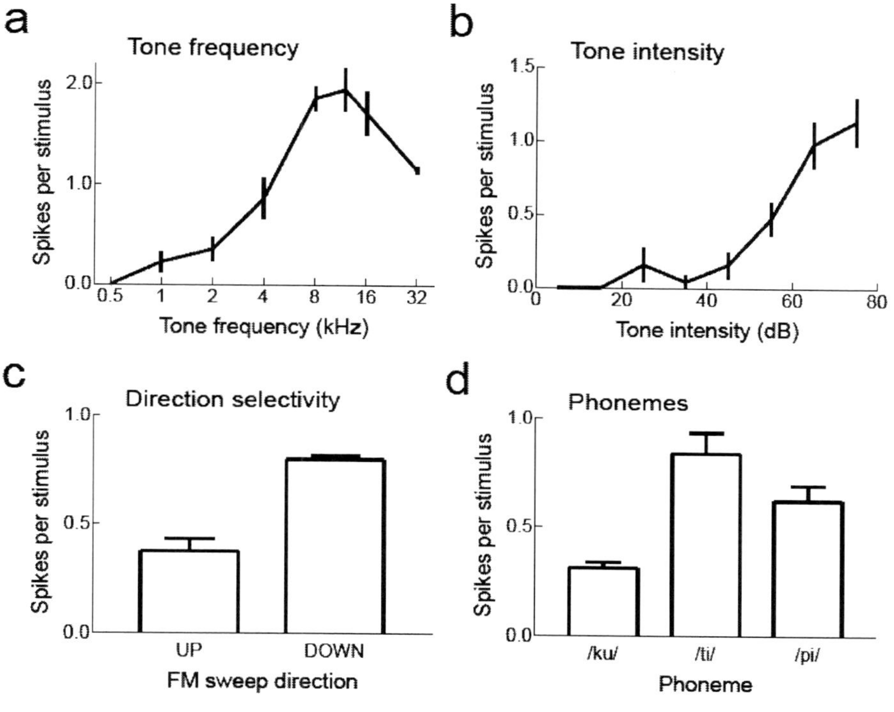

Figure 1. Cortical neurons are tuned to a wide range of stimulus features. a, Frequency (pitch) tuning in adult rat A1. The best frequency is 12 kHz. b, Intensity tuning in A1. The best level is 80 dB; note that this is the same recording as in a, demonstrating that individual cortical neurons can be selectively tuned across multiple variables. c, Direction selectivity of a rat A1 neuron for frequency-modulated (FM) sweeps. This neuron prefers downsweeps (32 kHz to 0.5 kHz) over upsweeps (0.5 kHz to 32 kHz). d, Phoneme preference of a primate A1 neuron. This cell selectively fired in response to /ti/ sounds.

Experiments using in vivo intracellular recording have obtained the highest resolution descriptions of receptive field composition. In these experiments, sharp electrodes or whole-cell electrodes have been used to measure the small sub- and suprathreshold excitatory and inhibitory inputs that dictate the spiking output of cortical neurons. Some of the initial findings came from cat V1 (Ferster & Lindstrom, 1983; Pei et al., 1991), showing that excitatory and inhibitory inputs seemed to be co-tuned for orientation tuning (Ferster, 1986). While the generality of co-tuned excitation and inhibition has been challenged in V1 (Douglas et al., 1991; Monier et al., 2003; Schummers et al., 2002), there is consensus in A1 that for synaptic tuning curves of frequency preference, excitation and inhibition are co-tuned and approximately balanced across the subthreshold frequency range (Volkov & Galazjuk, 1991; Wehr & Zador, 2003; Tan et al., 2004; Zhang et al., 2003). This agreement has made it possible to move beyond studies of the basic organization of synaptic receptive fields to ask questions regarding the plasticity of synaptic tuning curves (Froemke et al., 2007) and the disruption of synaptic tuning in nervous system disorders (Kenet et al., 2007).

CORTICAL RECEPTIVE FIELD PLASTICITY

Cortical representations are dynamic. Receptive field properties are constantly modified by experience and undergo fast reorganization during developmental critical periods. The critical period is an epoch in which the organization of cortical circuits is particularly susceptible to changes in sensory input. Critical periods generally last for a few days or weeks, beginning just after the onset of hearing or eye opening. The original studies of Wiesel and Hubel (1963) provided the initial reports of critical period plasticity in the visual system of kittens. In these early experiments, occlusion of one eye led to a dramatic shift in the ocular dominance of V1 neurons- cells that initially responded preferentially to input from the closed eye were re-tuned to respond to the open eye. In years since, monocular deprivation has remained a valuable tool for studying the mechanisms of cortical plasticity (Hensch, 2005; Singer et al., 1982). This is partially due to the relative simplicity in the organization of binocular vision- V1 is the first stage in which information from both eyes is integrated in single neurons, and thus there are only two inputs which compete for ocular dominance during the V1 critical period.

A series of studies from our own laboratory has documented the effects of patterned acoustic stimulation on development of rat A1. The auditory system is advantageous for studying how the precise statistics of sensory input determine the organization of cortical networks because of the high degree of acoustic stimulus control available, even for experiments with freely moving animals. We have systematically examined how passive exposure to various forms of spectrally- and temporally-varying auditory stimuli differentially shape A1 receptive fields and map organization. In particular, we have compared the results of exposure during early life to broadband (white noise) vs. narrowband (pure tone) stimuli, and we have also examined the consequences of stimulus presentation with tonic (continual) vs. phasic (short pulses) temporal patterns.

Cortical representations of sound frequency and intensity were both altered in a profound and persistent way if young rats were exposed to pulsed pure tones for a brief period between postnatal day 11 and 13, leading to rapid tonotopic map maturation and bringing the A1 critical period to an abrupt end (de Villers-Sidani et al., 2007). Complementary studies with pulsed white noise stimuli demonstrated that A1 frequency selectivity was degraded if rats were passively exposed to pulsed noise stimuli early in life (Zhou & Merzenich, 2007). Therefore, pure tone exposure and white noise exposure had opposite effects on the feature selectivity of A1 neurons; in both cases, however, receptive fields seem to be remodeled to match the frequency structure of the sensory environment, and the closure of the A1 critical period was accelerated. Exposure to continual white noise also degraded cortical receptive fields, while prolonging the extent of the critical period into adulthood (Chang & Merzenich, 2003). We conclude that the temporal pattern of sensory input regulates the closure of the A1 critical period- continuous stimuli keep the critical period open, perhaps because of the strong neuronal adaptation driven by tonic input, while pulsed or phasic input precociously close the critical period, likely due to the increase in correlated or coincidental activity between groups of cells and the consequent increase in long-term synaptic modifications occurring throughout the cortical network.

Cortical receptive fields remain plastic all throughout life, although the threshold for induction is likely to be higher in older brains. Animals previously exposed to white noise that underwent perceptual training later in life showed recovery of normal cortical representation, showing that training could normalize the degraded cortical responses and map organization in A1 (Zhou & Merzenich, 2007). In primary somatosensory cortex (S1) of adult monkeys, removal of normal peripheral input by amputation of the fifth digit resulted in gradual reorganization of the denervated cortical area, such that S1 receptive fields were retuned to prefer neighboring regions of the intact hand (Merzenich et al., 1983; Rasmusson, 1982). These functional alterations were presumably supported by disinhibition or potentiation of intracortical connections from nearby neurons or previously quiescent thalamic terminals, although historically it has been difficult to precisely determine which connections are predominantly responsible for receptive field modifications.

The specific organization and the dynamic aspects of receptive fields depend on the patterns of neuronal connectivity and the strength of these connections within and between different layers of cortex. Cortical neurons utilize feedforward, feedback, and recurrent synaptic connections to perform specialized processing functions, with the particular synaptic connectivity pattern determining the excitatory, inhibitory, plastic, and stable regions of the receptive field (Froemke et al., 2007; Schummers et al., 2004; Tan et al., 2004). Shifts in the spatial receptive field location or size can be quite extensive, and in some cases extend beyond the range of the primary thalamic input (Budinger et al., 2000; Carlson et al., 1986; Clark et al., 1988; Jenkins et al., 1990). These studies imply that alterations in receptive fields occur predominantly through physiological changes to specific inputs emerging from multiple cortical and subcortical sources, rather than changes in anatomical connectivity. However, over longer time periods and with more invasive alterations in sensory input (e.g., after peripheral lesions), anatomical changes have been observed (Darian-Smith & Gilbert, 1994).

The dynamics of cortical receptive fields and maps are also partially governed by the temporal properties of afferent inputs. For instance, altering the temporal correlation of inputs from two adjacent digits on the adult owl monkey hand resulted in enhancement of the correlation of inputs from skin surfaces of adjacent fingers (Clark et al., 1988). The timing specificity of receptive field modifications can be highly precise, on the order of hundreds or even tens of milliseconds (Fu et al., 2003; Jackson et al., 2006; Wang et al., 1995; Yao & Dan, 2001). This sensitivity is believed to be due to the temporal precision of the underlying synaptic modifications: the sign and magnitude of cortical synaptic plasticity is determined by the order and interval between pre- and postsynaptic spiking, a phenomenon dubbed spike-timing-dependent plasticity (STDP). STDP has been observed at several intracortical synapses (Dan & Poo, 2004; Feldman, 2000; Froemke & Dan, 2002; Markram et al., 1997; Sjöström et al., 2001), albeit with somewhat different learning rules depending on the particular type of synaptic connection (Froemke et al., 2005; Sjöström & Hausser, 2006). As a protocol for inducing long-term changes in synaptic strength, STDP is attractive as it offers an unprecedented degree of control over the exact amount of modification in a highly-predictable manner (Froemke et al., 2006), and has been used extensively in theoretical models and simulations of cortical function and organization (Froemke et al., 2005; Izhikevich, 2007; Song & Abbott, 2001).

Several experimental and theoretical studies have attempted to directly relate receptive field modifications to STDP (Allen et al., 2003; Fu et al., 2003; Shen et al., 2005; Song & Abbott, 2001; Yao & Dan, 2001). A recent study directly examined the effect of spike pairing on receptive field organization in anesthetized young rat V1 (Meliza & Dan, 2006). Pairing visual stimuli with postsynaptic spikes led to an expansion or contraction of excitatory receptive field size, with a time window matching that of STDP in young rat V1 brain slices (Froemke & Dan, 2002). Similarly, a set of studies in the developing tadpole visual system has demonstrated that sensory input paired with postsynaptic spiking led to changes in the synaptic receptive fields evoked by visual stimulation (Lien et al., 2006; Tao et al., 2001). However, spike pairing may be sufficient for induction of long-term synaptic receptive field modification only during development, as a study from our laboratory in adult rat A1 in vivo found that pairing tone-evoked responses with postsynaptic spikes had no enduring effects on synaptic strength (Froemke et al., 2007). These results are reminiscent of earlier findings in rat A1, in which repetitive but behaviorally-irrelevant auditory stimulation led to large changes in cortical map organization early in life, but were ineffective in adulthood (Zhang et al., 2001).

Cortical receptive fields are strongly modified during conditioning and perceptual learning. Classical conditioning, using pure tones of a given pitch as the conditioned stimulus, is sufficient to induce frequency-specific receptive-field plasticity in A1. This is observed as a selective enhancement of cortical responses to the frequency of the conditioned stimulus, as opposed to other frequencies or the best frequency (Weinberger et al., 1993). Classical conditioning paradigms have been used to map the changes in receptive fields in discrimination tasks of varying complexity. Guinea pigs trained to discriminate tones using foot shock as an unconditioned stimulus showed high behavioral performance on simple tasks that involved discrimination of negative conditioned stimulus from 30 intermixed trials of positive conditioned stimulus (Edeline & Weinberger, 1993). This was in sharp contrast to the behavioral performance on more complex tasks; animals showed poor performance if the distance between the frequencies of positively and negatively conditioned stimuli is reduced. Frequency-specific receptive field plasticity was observed in simple as well as complex tasks. These experiments revealed that the responses to the conditioned stimulus frequency are enhanced while the responses to other frequencies are reduced, usually at the best frequency. Reductions at the original best frequency are consistently observed in studies of receptive field plasticity induced in various systems and with different procedures (Bakin & Weinberger, 1996; Bakin et al., 1996; Bao et al., 2001; Froemke et al., 2007; Suga & Ma, 2003; Weinberger, 2004), indicating that cortical neurons have access to information about the overall structure of their tuning curves, and that wide-scale changes can be coordinated across the receptive field as a whole.

Receptive field plasticity also occurs during instrumental learning. In parallel to classical conditioning, long-term modifications develop when animals are exposed to either one-tone avoidance conditioning or two-tone discrimination training. Animals that are trained for avoidance conditioning paradigm show long-term receptive field modifications. A crucial factor is the suppression of fear responses during avoidance-conditioned training (Bakin et al., 1996). Likewise, rats can be trained to discriminate tones of variable frequency using a modified go/no-go training strategy. This task involves training noise-reared rats to identify target auditory stimuli of specific frequency from a set of distractors of varying frequencies. Intensive training during adulthood leads to improved performance in identifying selective target frequency among many distractors (Zhou & Merzenich, 2007), suggesting that training programs that capitalize on the mechanisms of perceptual learning might be effective in repairing damaged or disorganized human brains as well.

Neuromodulation and Receptive Field Plasticity

Receptive fields are controlled by changes to the pattern and timing of sensory inputs, in conjunction with neuromodulatory cues that signify behavioral context. Collective findings from several labs have determined that receptive field plasticity is governed by diverse subcortical structures that project to A1 and provide neuromodulatory input. Foremost among these is the nucleus basalis of the basal forebrain, the main source of acetylcholine in the cortex. Acetylcholine release is essential for learning and memory, and is believed to be involved in arousal and attentional modulation of cortical responses (Froemke et al., 2007; Parikh et al., 2007). An extensive number of studies using extracellular recordings have shown that pairing pure tones of a specific frequency with electrical stimulation of nucleus basalis ('nucleus basalis pairing') induces large, long-lasting enhancements of spiking (Bakin & Weinberger, 1996; Kilgard & Merzenich, 1998; Rasmusson & Dykes, 1988). These changes are robust across species and systems, and are observable at the level of field potentials, single- and multi-unit recordings, and cortical maps as a whole. Stimulation of the nucleus basalis is expected to activate a heterogeneous population of projection neurons, including those that release acetylcholine, GABA, and peptides; however, it is generally believed that cholinergic transmission- and the corresponding activation of cortical muscarinic acetylcholine receptors- mediates most of the long-term effects of this pairing procedure. Acetylcholine has a wide range of effects on cortical neurons, but a consistent observation is increased excitability (Woody & Gruen, 1987) and suppression of intracortical synaptic transmission (Metherate et al., 2005; Sarter & Parikh, 2005; Xiang et al., 1998). However, it has been unclear how these effects could produce long-term response enhancement specific for particular stimuli, such as pure tones of the paired frequency. As extracellular recordings cannot reveal the changes to synaptic inputs that are ultimately responsible for receptive field plasticity, we combined whole-cell recording with nucleus basalis stimulation in the intact brain to determine the synaptic basis of cortical receptive field plasticity (Figure 2). These experiments additionally serve as a model for understanding the basic principles of neuromodulation and synaptic plasticity in vivo, i.e., under more naturalistic conditions than those in brain slices.

To stimulate the nucleus basalis, electrodes were first implanted in this nucleus in adult rats. We then made in vivo whole-cell recordings from A1 neurons (Fig. 2a). For each cell, we first measured synaptic frequency tuning (Fig. 2b), and then repetitively paired tones of a specific non-preferred frequency with simultaneous tetanic stimulation of the nucleus basalis. During this nucleus basalis pairing procedure, there was a large decrease in the amplitude of inhibition evoked by the paired tone that occurred several seconds after pairing was initiated; this dramatic inhibitory LTD was followed approximately one minute later by delayed LTP of excitation, also specific to the tones of the paired frequency (Froemke et al., 2007). Long-term excitatory and inhibitory modifications induced by nucleus basalis pairing required co-activation of muscarinic acetylcholine receptors and NMDA receptors, indicating that a complex set of biochemical mechanisms were coordinated across multiple inputs to enable modifications of synaptic receptive fields. Careful examination of intracortical and thalamic inputs showed that these synaptic modifications to A1 neurons occurred almost entirely within intracortical connections and not to the primary thalamic input to A1 (Froemke et al., 2007).

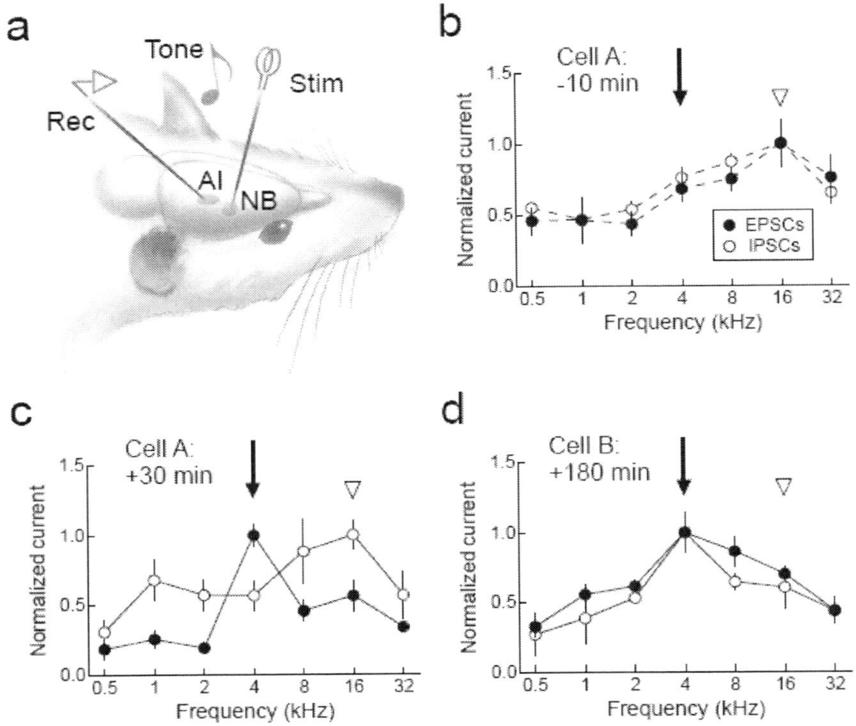

Figure 2. Temporal dynamics of progressive synaptic receptive field plasticity induced by nucleus basalis pairing. a, Experimental configuration. The stimulation electrode was acutely implanted in the right nucleus basalis, and whole-cell recordings were obtained from neurons in right A1. Pure tones of various frequencies were played to the contralateral ear, and synaptic responses were recorded in voltage-clamp. b, Frequency tuning of synaptic excitation (filled) and inhibition (open) for the first cell 10 minutes prior to nucleus basalis pairing. Arrow indicates the paired frequency (4 kHz). Arrowhead indicates the original best frequency (16 kHz) for this region of A1. Note the balance of excitation and inhibition across frequencies. c, Frequency tuning of the same cell in b, recorded 30 minutes after nucleus basalis pairing. The E:I ratio at the paired frequency has doubled, due to the enhancement of excitation and suppression of inhibition after nucleus basalis pairing. The paired frequency is the best frequency for excitation but not inhibition. d, Another cell from same region of A1, recorded 180 minutes after nucleus basalis pairing. The E:I ratio at the paired frequency is 1.0, as the paired frequency is the best frequency for both excitation and inhibition. Adapted from Froemke et al. (2007).

The inhibitory and excitatory synaptic modifications were long lasting: we found that nucleus basalis pairing led to a long-term increase in the ratio of excitation to inhibition (the 'E:I ratio'), as a consequence of the cooperative effects of suppression of inhibition and enhancement of excitation (Fig. 2c). The specific, enduring increase in E:I ratio was surprising, given the overall balance of excitation and inhibition for AI neurons before pairing (Fig. 2b). We noticed that towards the end of long-term recording sessions (up to one hour after pairing), inhibition at the paired frequency seemed to recover. Because of the technical difficulties in maintaining stable whole-cell recordings for longer than 30 minutes in vivo, we followed the temporal dynamics of synaptic receptive field plasticity by making several consecutive whole-cell recordings from the same location in AI for hours after nucleus basalis pairing. We found that the change in E:I ratio continued to evolve for hours after pairing, with inhibition eventually increasing to a higher level than before, to re-balance the persistent increase of excitation at the paired frequency (Fig. 2d).

These results indicate that the dynamics of inhibitory transmission could serve as a synaptic memory trace of the pairing event (Froemke et al., 2007). The duration of input-selective disinhibition permits self-reorganization of A1 tuning curves to emphasize the new preference for the paired stimulus, in a manner independent of further evoked neuromodulator release (e.g., compare Figs. 2b and 2d). Under natural conditions, this memory trace would represent episodes or stimuli that have acquired new behavioral meaning, or might be similar to the sorts of cortical changes that occur during perceptual learning, especially for those tasks requiring focal attention and sensory discrimination.

We hypothesize that the suppression of inhibition is a general mechanism for induction of receptive field modification in the adult cortex. During developmental critical periods, the high level of plasticity may be due to a less-refined inhibitory tone (Chang et al., 2005), permissive for alterations of cortical networks by passive stimuli. In adult cortex, however, active release of neuromodulators is required for extensive receptive field plasticity, reflecting the importance of behavioral context in associative learning and memory provided by subcortical systems (Doya et al., 2002). This is further demonstrated by an excellent series of studies from Fritz and colleagues (Fritz et al., 2003; Fritz et al., 2005). Single-unit recordings of A1 receptive fields were obtained from head-restrained behaving ferrets. The receptive fields of A1 neurons were found to be powerfully modified by behavioral conditioning procedures. Excitatory and suppressive subregions of extracellular receptive fields evoked by certain auditory stimuli were altered when those stimuli were followed by tail-shock. Strikingly similar to the synaptic effects of nucleus basalis pairing (Froemke et al., 2007), the most common observations were increases of excitatory regions and reductions of suppressive regions of the receptive field around the conditioned tone (Fritz et al., 2003). These changes in receptive field structure could endure for minutes to hours after conditioning, possibly serving as a memory in sensory cortex for the contingencies of the training regime.

An important aspect of the neuromodulatory systems is that they seem to be involved in fine-tuning of cortical responses, enabling animals to attend to behaviorally relevant stimuli and ignore irrelevant ones. These neuromodulatory systems enable cortical plasticity by directing the cortex to selectively respond to important or novel stimuli, possibly by disinhibition of local cortical circuits. The combination of electrical and acoustic stimulation can change several properties of receptive fields, contingent on the details of the sensory input. For instance, nucleus basalis pairing has been observed to narrow or broaden receptive field size, depending on the number and range of tones presented during pairing (Kilgard et al., 2001).

Neuromodulators besides acetylcholine can also affect cortical receptive fields and map organization. While there are many neuromodulators and hormones that can alter cortical function, one prominent system is the dopaminergic projection from the VTA. Midbrain dopamine neurons are activated by novel stimuli or unpredicted rewards, and seem to provide reinforcement signals during learning and conditioning procedures (Schultz, 2007). Similar to the nucleus basalis pairing procedure described above, pairing auditory stimuli with electrical stimulation of the VTA has been found to result in cortical map reorganization, such that the cortical area devoted to paired tones are specifically changed (Bao et al., 2001). Repetitive VTA pairing can induce either increases or decreases in cortical map area, depending on the temporal relationship between the paired tone and VTA stimulation, much like the timing dependencies observed in STDP and classical conditioning. When the tone consistently led VTA stimulation, such that it effectively predicted VTA activity, the representation of that stimulus was enhanced. Conversely, when the tone consistently followed VTA stimulation, stimulus representation was reduced. In contrast to the effects of nucleus basalis pairing, VTA signaling provides a mechanism for bidirectional control of cortical network activity. However, as A1 itself receives minimal direct input from the VTA, some of the largest effects of VTA pairing were observed in auditory cortical areas outside of A1. This suggests that most of the changes to A1 maps induced by VTA pairing are indirect consequences of map plasticity in higher cortical areas, presumably communicated by intracortical connections. Studies in brain slices indicate that direct modulation of NMDA receptors may be the primary means through which dopamine release affects synaptic strength (Seamans & Yang, 2004). While nucleus basalis stimulation is seemingly different, acting directly on GABAergic transmission instead (Froemke et al., 2007), the net effect may be the same: regulation of calcium signaling by NMDA receptors that ultimately controls the sign and magnitude of long-term synaptic modifications (Froemke et al., 2005; Malenka & Bear, 2004).

Neuromodulation of synaptic plasticity and learning is a general feature of cortical networks. For example, receptive fields of S1 neurons can be modified by pairing somatosensory stimulation with nucleus basalis activation (Alenda & Nunez, 2007). In V1, pairing a visual stimulus with iontophoretic application of the neuromodulators such as acetylcholine, noradrenaline or glutamate resulted in persistent, selective changes in the responses to the paired feature (Greuel et al., 1988). Besides animal models, cortical plasticity has also been observed in human motor cortex (M1). Using transcranial magnetic stimulation (TMS) and electromyography recording, small motor evoked potentials (MEPs) can be measured in human subjects. MEP size can be persistently changed both by practice and repetitive TMS to the corresponding region of M1 (Meintzschel & Ziemann, 2006). Intriguingly, MEP modifications require the activation of a number of neuromodulator systems, including the cholinergic, dopaminergic, and noradrenergic systems, as specific agonists and antagonists can enhance or block the changes to MEPs, respectively. Neuromodulation and plasticity are thus important mechanisms for analysis of sensory information, even at the earliest stages of cortical processing, in humans as well as animals.

Implications of Cortical Plasticity to Overall Brain Function

These studies of cortical receptive field plasticity have provided insight into the process by which information about the sensory world is encoded and updated by specific neural circuits. The capacity of cortex to rapidly reorganize allows cortical neurons and ensembles to respond to novel or behaviorally important stimuli with high fidelity, and to dynamically readjust the emphasis on different stimuli depending on the current behavioral context and the previous history of the organism. Thus it seems that cortical plasticity is important for cognition, memory, motor control, attention, perception and executive functions through out the life span of a human or an animal. The progressive decline of information processing in aging rats and humans has been attributed to degradation of receptive field organization and plastic processes (Byl et al., 1996; Strata et al., 2005; Zhang et al., 2002; Zhou & Merzenich, 2007). Impaired inhibitory transmission, noise generation, weakened neuromodulatory inputs, and loss of learning capabilities through synapse elimination may be some of the prime factors that may contribute to impaired or detrimental plasticity. Fortunately, it is possible that behavioral manipulations can be used as an effective strategy to restore cortical plasticity and the progressive loss of cortical processing abilities throughout aging.

Breakdown in cortical receptive field organization and plasticity is also postulated to play a role in neurodevelopmental disorders such as fragile-x and autism, as well as speech and language disorders such as dyslexia. Although genetic variation may be a major factor for these disorders, social, emotional and behavioral alterations are precisely guided by processes and mechanisms that control cortical plasticity. Recently we have proposed a model involving an imbalance in the ratio of excitation to inhibition that could help explain autistic traits (Rubenstein & Merzenich, 2003). Cortical plasticity may also important for controlling and coordinating language abilities such as speech processing and speech production. The progressive decline in reading abilities observed in dyslexia is also partially attributed to changes in the organization of sensory cortex that in-turn affect sensory-motor integration (Merzenich et al., 1993).

Cortical plasticity thus spans small-scale synaptic changes, mostly understood in terms of the underlying electrical and molecular mechanisms, to larger-scale changes over neural circuits that directly govern behavior. Receptive field properties are determined by behavioral experience, attention, and motivation, via the spiking statistics and neuromodulatory input received by local cortical networks. These changes have been documented for much of cortex, including visual, auditory, somatosensory and motor cortical areas in animal models and humans. Synaptic modifications induced in these cortical areas are hypothesized to be the basis of perceptual learning, and gradually, we are beginning to understand the neural mechanisms of learning in terms of the basic synaptic elements. One essential component of learning is neuromodulation. A thorough description of the effects of different neuromodulators, from the synaptic to cognitive levels, will be required to understand the relationships between the environment, the brain, cognition, and behavior.

ACKNOWLEDGMENTS

This work was supported by grants from the NIDCD, the Jane Coffin Childs Foundation for Medical Research, and the Sandler Foundation.

REFERENCES

Alenda, A., & Nunez, A. (2007). Cholinergic modulation of sensory interference in rat primary somatosensory cortical neurons. *Brain Res, 1133*(1), 158-167.

Allen, C. B., Celikel, T., & Feldman, D. E. (2003). Long-term depression induced by sensory deprivation during cortical map plasticity in vivo. *Nat Neurosci, 6*(3), 291-299.

Bakin, J. S., South, D. A., & Weinberger, N. M. (1996). Induction of receptive field plasticity in the auditory cortex of the guinea pig during instrumental avoidance conditioning. *Behav Neurosci, 110*(5), 905-913.

Budinger, E., Heil, P., & Scheich, H. (2000). Functional organization of auditory cortex in the Mongolian gerbil (Meriones unguiculatus). III. Anatomical subdivisions and corticocortical connections. *Eur J Neurosci, 12*(7), 2425-2451.

Buonomano, D. V., & Merzenich, M. M. (1998). Cortical plasticity: from synapses to maps. *Annu Rev Neurosci, 21*, 149-186.

Byl, N. N., Merzenich, M. M., & Jenkins, W. M. (1996). A primate genesis model of focal dystonia and repetitive strain injury: I. Learning-induced dedifferentiation of the representation of the hand in the primary somatosensory cortex in adult monkeys. Neurology, 47(2), 508-520.

Carlson, M., Huerta, M. F., Cusick, C. G., & Kaas, J. H. (1986). Studies on the evolution of multiple somatosensory representations in primates: the organization of anterior parietal cortex in the New World Callitrichid, Saguinus. *J Comp Neurol, 246*(3), 409-426.

Chang, E. F., & Merzenich, M. M. (2003). Environmental noise retards auditory cortical development. *Science, 300*(5618), 498-502.

Chang, E. F., Bao, S., Imaizumi, K., Schreiner, C. E., & Merzenich, M. M. (2005). Development of spectral and temporal response selectivity in the auditory cortex. *Proc Natl Acad Sci U S A, 102*(45), 16460-16465.

Clark, S. A., Allard, T., Jenkins, W. M., & Merzenich, M. M. (1988). Receptive fields in the body-surface map in adult cortex defined by temporally correlated inputs. *Nature, 332*(6163), 444-445.

Dan, Y., & Poo, M. M. (2004). Spike timing-dependent plasticity of neural circuits. *Neuron, 44*(1), 23-30.

Darian-Smith, C., & Gilbert, C. D. (1994). Axonal sprouting accompanies functional reorganization in adult cat striate cortex. *Nature, 368*(6473), 737-740.

de Villers-Sidani, E., Chang, E. F., Bao, S., & Merzenich, M. M. (2007). Critical period window for spectral tuning defined in the primary auditory cortex (A1) in the rat. *J Neurosci, 27*(1), 180-189.

Douglas, R. J., Martin, K. A., & Whitteridge, D. (1991). An intracellular analysis of the visual responses of neurones in cat visual cortex. *J Physiol, 440*, 659-696.

Doya, K., Samejima, K., Katagiri, K., & Kawato, M. (2002). Multiple model-based reinforcement learning. *Neural Comput, 14*(6), 1347-1369.

Edeline, J. M., & Weinberger, N. M. (1993). Receptive field plasticity in the auditory cortex during frequency discrimination training: selective retuning independent of task difficulty. *Behav Neurosci, 107*(1), 82-103.

Feldman, D. E. (2000). Timing-based LTP and LTD at vertical inputs to layer II/III pyramidal cells in rat barrel cortex. *Neuron, 27*(1), 45-56.

Ferster, D. (1986). Orientation selectivity of synaptic potentials in neurons of cat primary visual cortex. *J Neurosci, 6*(5), 1284-1301.

Ferster, D., & Lindstrom, S. (1983). An intracellular analysis of geniculo-cortical connectivity in area 17 of the cat. *J Physiol, 342*, 181-215.

Fritz, J., Shamma, S., Elhilali, M., & Klein, D. (2003). Rapid task-related plasticity of spectrotemporal receptive fields in primary auditory cortex. *Nat Neurosci, 6*(11), 1216-1223.

Fritz, J. B., Elhilali, M., & Shamma, S. A. (2005). Differential dynamic plasticity of A1 receptive fields during multiple spectral tasks. *J Neurosci, 25*(33), 7623-7635.

Froemke, R. C., & Dan, Y. (2002). Spike-timing-dependent synaptic modification induced by natural spike trains. *Nature, 416*(6879), 433-438.

Froemke, R. C., Merzenich, M. M., & Schreiner, C. E. (2007). A synaptic memory trace for cortical receptive field plasticity. *Nature, 450*(7168), 425-429.

Froemke, R. C., Poo, M. M., & Dan, Y. (2005). Spike-timing-dependent synaptic plasticity depends on dendritic location. *Nature, 434*(7030), 221-225.

Froemke, R. C., Tsay, I. A., Raad, M., Long, J. D., & Dan, Y. (2006). Contribution of individual spikes in burst-induced long-term synaptic modification. *J Neurophysiol, 95*(3), 1620-1629.

Fu, Y.-X., Djupsund, K., Gao, H., Hayden, B., Shen, K., & Dan, Y. (2002). Temporal specificity in the cortical plasticity of visual space representation. *Science, 296*, 1999-2003.

Greuel, J. M., Luhmann, H. J., & Singer, W. (1988). Pharmacological induction of use-dependent receptive field modifications in the visual cortex. *Science, 242*(4875), 74-77.

Gu, Q. (2002). Neuromodulatory transmitter systems in the cortex and their role in cortical plasticity. *Neuroscience, 111*(4), 815-835.

Hensch, T. K. (2005). Critical period plasticity in local cortical circuits. *Nat Rev Neurosci, 6*(11), 877-888.

Imaizumi, K., Priebe, N. J., Crum, P. A., Bedenbaugh, P. H., Cheung, S. W., & Schreiner, C. E. (2004). Modular functional organization of cat anterior auditory field. *J Neurophysiol, 92*(1), 444-457.

Izhikevich, E. M. (2007). Solving the distal reward problem through linkage of STDP and dopamine signaling. *Cereb Cortex, 17*(10), 2443-2452.

Jackson, A., Mavoori, J., & Fetz, E. E. (2006). Long-term motor cortex plasticity induced by an electronic neural implant. *Nature, 444*(7115), 56-60.

Jenkins, W. M., Merzenich, M. M., Ochs, M. T., Allard, T., & Guic-Robles, E. (1990). Functional reorganization of primary somatosensory cortex in adult owl monkeys after behaviorally controlled tactile stimulation. *J Neurophysiol, 63*(1), 82-104.

Karmarkar, U. R., & Dan, Y. (2006). Experience-dependent plasticity in adult visual cortex. *Neuron, 52*(4), 577-585.

Katz, L. C., & Shatz, C. J. (1996). Synaptic activity and the construction of cortical circuits. *Science, 274*(5290), 1133-1138.

Kenet, T., Froemke, R. C., Schreiner, C. E., Pessah, I. N., & Merzenich, M. M. (2007). Perinatal exposure to a noncoplanar polychlorinated biphenyl alters tonotopy, receptive fields, and plasticity in rat primary auditory cortex. *Proc Natl Acad Sci U S A, 104*(18), 7646-7651.

Kilgard, M. P., & Merzenich, M. M. (1998). Cortical map reorganization enabled by nucleus basalis activity. *Science, 279*(5357), 1714-1718.

Kilgard, M. P., Pandya, P. K., Vazquez, J., Gehi, A., Schreiner, C. E., & Merzenich, M. M. (2001). Sensory input directs spatial and temporal plasticity in primary auditory cortex. *J Neurophysiol, 86*(1), 326-338.

Lien, C. C., Mu, Y., Vargas-Caballero, M., & Poo, M. M. (2006). Visual stimuli-induced LTD of GABAergic synapses mediated by presynaptic NMDA receptors. *Nat Neurosci, 9*(3), 372-380.

Liu, R. C., & Schreiner, C. E. (2007). Auditory cortical detection and discrimination correlates with communicative significance. *PLoS Biol, 5*(7), e173.

Malenka, R. C., & Bear, M. F. (2004). LTP and LTD: an embarrassment of riches. *Neuron, 44*(1), 5-21.

Markram, H., Lubke, J., Frotscher, M., & Sakmann, B. (1997). Regulation of synaptic efficacy by coincidence of postsynaptic APs and EPSPs. *Science, 275*(5297), 213-215.

Meintzschel, F., & Ziemann, U. (2006). Modification of practice-dependent plasticity in human motor cortex by neuromodulators. *Cereb Cortex, 16*(8), 1106-1115.

Meliza, C. D., & Dan, Y. (2006). Receptive-field modification in rat visual cortex induced by paired visual stimulation and single-cell spiking. *Neuron, 49*(2), 183-189.

Merzenich, M. M., Kaas, J. H., Wall, J. T., Sur, M., Nelson, R. J., & Felleman, D. J. (1983). Progression of change following median nerve section in the cortical representation of the hand in areas 3b and 1 in adult owl and squirrel monkeys. *Neuroscience, 10*(3), 639-665.

Merzenich, M. M., Schreiner, C. E., Jenkins, W., & Wang, X. (1993). Neural mechanisms underlying temporal integration, segmentation, and input sequence representation: some implications for the origin of learning disabilities. *Ann N Y Acad Sci, 682*, 1-22.

Metherate, R., Kaur, S., Kawai, H., Lazar, R., Liang, K., & Rose, H. J. (2005). Spectral integration in auditory cortex: mechanisms and modulation. *Hear Res, 206*(1-2), 146-158.

Monier, C., Chavane, F., Baudot, P., Graham, L. J., & Fregnac, Y. (2003). Orientation and direction selectivity of synaptic inputs in visual cortical neurons: a diversity of combinations produces spike tuning. *Neuron, 37*(4), 663-680.

Parikh, V., Kozak, R., Martinez, V., & Sarter, M. (2007). Prefrontal acetylcholine release controls cue detection on multiple timescales. *Neuron, 56*(1), 141-154.

Pei, X., Volgushev, M., Vidyasagar, T. R., & Creutzfeldt, O. D. (1991). Whole cell recording and conductance measurements in cat visual cortex in-vivo. *Neuroreport, 2*(8), 485-488.

Rasmusson, D. D. (1982). Reorganization of raccoon somatosensory cortex following removal of the fifth digit. *J Comp Neurol, 205*(4), 313-326.

Rasmusson, D. D., & Dykes, R. W. (1988). Long-term enhancement of evoked potentials in cat somatosensory cortex produced by co-activation of the basal forebrain and cutaneous receptors. *Exp Brain Res, 70*(2), 276-286.

Rubenstein, J. L., & Merzenich, M. M. (2003). Model of autism: increased ratio of excitation/inhibition in key neural systems. *Genes Brain Behav, 2*(5), 255-267.

Safran, A. B., & Landis, T. (1996). Plasticity in the adult visual cortex: implications for the diagnosis of visual field defects and visual rehabilitation. *Curr Opin Ophthalmol, 7*(6), 53-64.

Sarter, M., & Parikh, V. (2005). Choline transporters, cholinergic transmission and cognition. *Nat Rev Neurosci, 6*(1), 48-56.

Schreiner, C. E., & Winer, J. A. (2007). Auditory cortex mapmaking: principles, projections, and plasticity. *Neuron, 56*(2), 356-365.

Schultz, W. (2007). Multiple dopamine functions at different time courses. *Ann Rev Neurosci, 30*, 259-288.

Schummers, J., Marino, J., & Sur, M. (2002). Synaptic integration by V1 neurons depends on location within the orientation map. *Neuron, 36*(5), 969-978.

Schummers, J., Marino, J., & Sur, M. (2004). Local networks in visual cortex and their influence on neuronal responses and dynamics. *J Physiol Paris, 98*(4-6), 429-441.

Seamans, J. K., & Yang, C. R. (2004). The principal features and mechanisms of dopamine modulation in the prefrontal cortex. *Prog Neurobiol, 74*(1), 1-58.

Shen, Y. S., Gao, H., & Yao, H. (2005). Spike timing-dependent synaptic plasticity in visual cortex: a modeling study. *J Comput Neurosci, 18*(1), 25-39.

Singer, W., Tretter, F., & Yinon, U. (1982). Evidence for long-term functional plasticity in the visual cortex of adult cats. *J Physiol, 324*, 239-248.

Sjöström, P. J., & Hausser, M. (2006). A cooperative switch determines the sign of synaptic plasticity in distal dendrites of neocortical pyramidal neurons. *Neuron, 51*(2), 227-238.

Sjöström, P. J., Turrigiano, G. G., & Nelson, S. B. (2001). Rate, timing, and cooperativity jointly determine cortical synaptic plasticity. *Neuron, 32*(6), 1149-1164.

Song, S., & Abbott, L. F. (2001). Cortical development and remapping through spike timing-dependent plasticity. *Neuron, 32*(2), 339-350.

Strata, F., deIpolyi, A. R., Bonham, B. H., Chang, E. F., Liu, R. C., Nakahara, H., et al. (2005). Perinatal anoxia degrades auditory system function in rats. *Proc Natl Acad Sci U S A, 102*(52), 19156-19161.

Suga, N., & Ma, X. (2003). Multiparametric corticofugal modulation and plasticity in the auditory system. *Nat Rev Neurosci, 4*(10), 783-794.

Tan, A. Y., Atencio, C. A., Polley, D. B., Merzenich, M. M., & Schreiner, C. E. (2007). Unbalanced synaptic inhibition can create intensity-tuned auditory cortex neurons. *Neuroscience, 146*(1), 449-462.

Tan, A. Y., Zhang, L. I., Merzenich, M. M., & Schreiner, C. E. (2004). Tone-evoked excitatory and inhibitory synaptic conductances of primary auditory cortex neurons. *J Neurophysiol, 92*(1), 630-643.

Tao, H. W., Zhang, L. I., Engert, F., & Poo, M. (2001). Emergence of input specificity of ltp during development of retinotectal connections in vivo. *Neuron, 31*(4), 569-580.

Volkov, I. O., & Galazjuk, A. V. (1991). Formation of spike responses to sound tones in cat auditory cortex neurons: interaction of excitatory and inhibitory effects. *Neuroscience, 43*(2-3), 307-321.

Wang, X., Merzenich, M. M., Sameshima, K., & Jenkins, W. M. (1995). Remodelling of hand representation in adult cortex determined by timing of tactile stimulation. *Nature, 378*(6552), 71-75.

Wehr, M., & Zador, A. M. (2003). Balanced inhibition underlies tuning and sharpens spike timing in auditory cortex. *Nature, 426*(6965), 442-446.

Weinberger, N. M. (2004). Specific long-term memory traces in primary auditory cortex. *Nat Rev Neurosci, 5*(4), 279-290.

Weinberger, N. M., Javid, R., & Lepan, B. (1993). Long-term retention of learning-induced receptive-field plasticity in the auditory cortex. *Proc Natl Acad Sci U S A, 90*(6), 2394-2398.

Wiesel, T. N., & Hubel, D. H. (1963). Single-cell responses in striate cortex of kittens deprived of vision in one eye. *J Neurophysiol, 26*, 1003-1017.

Winer, J. A., & Schreiner, C. E., eds. (2005). The Inferior Colliculus. Springer.

Woody, C. D., & Gruen, E. (1987). Acetylcholine reduces net outward currents measured in vivo with single electrode voltage clamp techniques in neurons of the motor cortex of cats. *Brain Res, 424*(1), 193-198.

Xiang, Z., Huguenard, J. R., & Prince, D. A. (1998). Cholinergic switching within neocortical inhibitory networks. *Science, 281*(5379), 985-988.

Yao, H., & Dan, Y. (2001). Stimulus timing-dependent plasticity in cortical processing of orientation. *Neuron, 32*, 315-323.

Zhang, L. I., Bao, S., & Merzenich, M. M. (2001). Persistent and specific influences of early acoustic environments on primary auditory cortex. *Nat Neurosci, 4*(11), 1123-1130.

Zhang, L. I., Bao, S., & Merzenich, M. M. (2002). Disruption of primary auditory cortex by synchronous auditory inputs during a critical period. *Proc Natl Acad Sci U S A, 99*(4), 2309-2314.

Zhang, L. I., Tan, A. Y., Schreiner, C. E., & Merzenich, M. M. (2003). Topography and synaptic shaping of direction selectivity in primary auditory cortex. *Nature, 424*(6945), 201-205.

Zhou, X., & Merzenich, M. M. (2007). Intensive training in adults refines A1 representations degraded in an early postnatal critical period. *Proc Natl Acad Sci U S A, 104*(40), 15935-15940.

In:Cognitive Science Compendium Volume 2
Editors: Miao-Kun Sun

ISBN: 978-1-60456-775-5

Chapter 7

LINKS BETWEEN DOWN SYNDROME AND ALZHEIMER'S DISEASE

Sul-Hee Chung* and Woo-Joo Song

Graduate Program in Neuroscience, Institute for Brain Science and Technology, Inje University, 633-146 Gaegeum 2-dong, Busanjin-gu, Busan, South Korea 614-735

ABSTRACT

Individuals with Down syndrome (DS) show an early-onset form of Alzheimer's disease (AD). The brains of both DS and AD patients have the pathological hallmarks of AD of amyloid plaques and neurofibrillary tangles, which are insoluble deposits that consist of β-amyloid (Aβ) and hyperphosphorylated Tau, respectively. Endosomal abnormalities also occur invariably in the neurons of AD and DS brains before classical AD neuropathology begins to develop. Although the cause of this pathology in DS patients is not clearly understood, one potential mechanism is the overexpression of genes located on chromosome 21. Genes that are potentially responsible for this include β-amyloid precursor protein (APP) and dual-specificity tyrosine-(Y)-phosphorylation regulated kinase 1A (DYRK1A). Emerging evidence suggests the roles of DYRK1A in Aβ production and Tau hyperphosphorylation, implying that DYRK1A may be involved in the early-onset of AD in DS patients. The purpose of this review is to describe recent studies regarding the links between DS and AD, especially those emphasizing the role of DYRK1A-mediated phosphorylation.

Keywords: Down syndrome, Alzheimer's disease, DYRK1A, APP, amyloid plaques, neurofibrillary tangles.

* Address correspondence and reprint requests to Sul-Hee Chung, Tel.: 82-51-892-4185; Fax: 82-51-892-0059; e-mail: sulchung@inje.ac.kr or Woo-Joo Song, Tel.: 82-51-892-4186; Fax: 82-51-892-0059; e-mail: wjsong@inje.ac.kr

INTRODUCTION

Down syndrome (DS, Trisomy 21) is the most common genetic disorder, with a frequency of 1 in every 800 live births; moreover, this frequency is significantly rising with the increase age of women over 35 who give birth. DS is caused by the presence of an extra copy of all or part of human chromosome 21 (Jacobs et al., 1959; Lejeune et al., 1959). Individuals with DS exhibit mild to severe mental/cognitive deficit, which is a major factor in preventing these patients from leading fully independent lives (Korenberg et al., 1994; Patterson 1987; Pulsifer 1996). In addition, DS individuals show a variety of other anomalies, including low muscle tone (hypotonia), congenital heart defects, gastrointestinal malformations, immune and endocrine system defects, a high incidence of leukemia, and an early-onset of Alzheimer's disease (AD) (Korenberg et al., 1994).

As a result of improvements in living conditions and advances in medical care, individuals with DS can be expected to live up to the end of their sixth decade and, in some cases, beyond. Adults with DS, however, typically show premature aging and a decline in memory ability. In addition, they are particularly vulnerable to AD (Burger et al., 1973; Olson et al., 1969; Wisniewski et al., 1985). Nearly all individuals with DS develop the neuropathological lesions seen in AD by the age of 40 and become demented (Nixon 2005). DS patients also have pathological hallmarks in the form of amyloid plaques and neurofibrillary tangles (NFTs) that are indistinguishable from those of AD patients (Grundke-Iqbal et al., 1986; Hanger et al., 1991; Masters et al., 1985).

Recently a number of researchers have found that overexpression of DYRK1A in DS brains, resulting from triplication of the DYRK1A gene on human chromosome 21, may constitute the underlying cause of early-onset AD through the phosphorylation of microtubule-associated protein Tau and β-amyloid precursor protein (APP) (Kimura et al., 2007; Park et al., 2007; Ryoo et al., 2008; Ryoo et al., 2007). In this review, recent reports concerning the potential mechanisms linking the gap between DS and AD are summarized with a focus on the emerging roles of DYRK1A on early-onset AD pathogenesis in DS brains.

PATHOLOGICAL FEATURES OF AD IN DS PATIENTS

The most distressing symptom for adult DS patients over 40 years old is early-onset AD (Burger et al., 1973; Olson et al., 1969; Wisniewski et al., 1985). AD is an irreversible, age-related, progressive brain disorder characterized by dementia. The brains of both DS and AD patients have the pathological hallmarks of amyloid plaques and NFTs, which are insoluble deposits of β-amyloid (Aβ) and hyperphosphorylated Tau, respectively (Grundke-Iqbal et al., 1986; Iwatsubo et al., 1995; Mann 1988; Masters et al., 1985; Wisniewski et al., 1985).

The generation of Aβ peptides is initiated by cleavage of APP by the β-secretase enzyme, which is also known as BACE1 (beta-site amyloid precursor protein cleaving enzyme 1), and subsequently cleaved by the γ-secretase complex, which consists of at least four components: Presenilin, Nicastrin, Anterior pharynx defective phenotype 1, and Presenilin enhancer 2 (De Strooper, 2003). The importance of Aβ in AD pathogenesis is clearly shown in rare familial AD cases in which mutations in *APP* or *Presenilin* genes increase the production of Aβ.

However, a majority of AD patients fall into the sporadic late-onset category characterized by the lack of a known genetic basis of the disorder. Moreover, in these cases, the mechanisms that underlie the pathological changes are not well understood.

NFTs, another hallmark of AD, are comprised of paired helical filaments (PHFs) that are mainly composed of hyperphosphorylated Tau protein (Hanger et al., 1991). More than 30 known Tau phosphorylation sites and 7~10 phosphorylations per Tau protein have been observed in PHFs (Kopke et al., 1993; Ksiezak-Reding et al., 1992). Although Tau protein is phosphorylated by several kinases *in vitro*, the true number of kinases that are actually involved in Tau phosphorylation *in vivo* is unknown. Currently, only glycogen synthase kinase 3β, cyclin-dependent kinase 5, Dyrk1A, cAMP-dependent protein kinase and microtubule-affinity-regulating kinase are known to modulate the proline-rich regions of Tau phosphorylation *in vivo* at some level, either directly or indirectly (Johnson et al., 2004; Ryoo et al., 2007). Such phosphorylations can induce conformational changes of Tau from its less-phosphorylated, microtubule-bound form to its hyperphosphorylated, self-aggregated form (Bielska et al., 2006).

In addition to the two pathological hallmarks of AD, amyloid plaques and NFTs, growing evidence suggests a link between abnormal neuronal endosomal function and AD pathology. Neuronal endosome enlargement, reflecting altered endocytic function, is one of the earliest known neuropathological changes in AD. It was reported that several decades prior to the development of classical AD neuropathology, early endosomes become significantly enlarged in pyramidal neurons in a DS fetus as early as 28 weeks (Cataldo et al., 2000). Early endosomes are a major site of APP processing and a likely place for pathogenic Aβ production (Tate et al., 2006). The emergence of enlarged rab5-positive endosomes coincides with an initial increase in Aβ40 and Aβ42 peptides in ELISA and immunocytochemistry using numerous Aβ antibodies in neurons from AD brains, supporting that the endosomal pathology contributes significantly to Aβ overproduction and accumulation in sporadic AD and DS with AD (Cataldo et al., 2004). The Ts65Dn mice, the best-known genetic model of DS, carry a part of chromosome 16 of the mouse that is homologous to human chromosome 21; this also develops enlarged neuronal early endosomes (Cataldo et al., 2003). The Ts65Dn mice show increased immunoreactivity for several endosome fusion and recycling markers, including rab5, early endosomal antigen 1, and rab4 (Cataldo et al., 2003).

Candidate Genes on Chromosome 21 that Link DS and AD

APP gene.

An integral membrane protein APP consists of a large N-terminal extracellular domain, a single transmembrane domain, and a short cytoplasmic C-terminal domain. The tissue distributions of three APP isoforms are as follows: the APP695 form is predominantly expressed in neuronal cells, whereas the APP751 and APP770 forms are detected in all examined tissues (Kosik, 1992). The early onset and high incidence of AD in DS individuals have long been ascribed to the presence of an extra copy of the *APP* gene on chromosome 21. Recently, it was shown that the presence of an extra copy of APP with familial Alzheimer's disease (FAD) is associated with both APP overexpression and early-onset AD (Rovelet-Lecrux et al., 2006).

It was reported that triplication of the *APP* gene in Ts65Dn mice and individuals with DS is critical for the development of endocytic abnormalities (Cataldo et al., 2003). Selective deletion of one *APP* gene copy or a region containing *APP* from Ts65Dn mice eliminates the abnormal endocytic phenotype. In addition, Prasher *et al.* reported a clinical case of a DS patient who had a partial trisomy 21 without *APP* triplication; this patient reached the age of 78 without AD neuropathological symptoms (Prasher et al., 1998). These results establish a direct relationship between the *APP* gene and endosome dysfunction. However, APP-overexpressing mice did not show the endosome phenotype, implying that APP acts in concert with products of one or more additional genes on the chromosome 16 triplicated region (Cataldo et al., 2003).

DYRK1A gene, a potential linker between DS and AD.

The *DYRK1A* gene is localized on the Down syndrome critical region of human chromosome 21 and encodes a proline-directed protein kinase that may be responsible for DS mental retardation (Guimera et al., 1996; Kentrup et al., 1996; Shindoh et al., 1996; Song et al., 1996). DYRK1A is a mammalian ortholog of the *Drosophila* minibrain (*Mnb*) gene that is essential for normal postembryonic neurogenesis (Tejedor et al., 1995). DYRK1A is ubiquitously expressed with strong expression in the brain (Guimera et al., 1996; Hammerle et al., 2003; Marti et al., 2003; Rahmani et al., 1998; Song et al., 1996), and the levels of DYRK1A mRNA and protein have been shown to be overexpressed in DS brains (Dowjat et al., 2007; Guimera et al., 1999; Ryoo et al., 2008). DYRK1A has dual substrate specificities; autophosphorylation for self-activation takes place on the tyrosine-321 residue in the active loop of the catalytic domain, and proline-directed phosphorylation of target proteins takes place on serine/threonine residues (Kentrup et al., 1996; Lochhead et al., 2005). Dyrk1A potentially phosphorylates or interacts with a variety of proteins, including transcription factors (NFATc, STAT3, Forkhead, CREB, Gli1), splicing-related factors (SF3b1/SAP155, cyclin L2), proteins involved in synaptic functions (Dynamin 1, Synaptojanin 1, Amphiphysin 1, α-Synuclein), eukaryotic initiation factor 2Bε, Hip-1, Tau and APP, implying the multiple biological functions of DYRK1A (Galceran et al., 2003; Ryoo et al., 2008; Ryoo et al., 2007).

One of the hypothesized DYRK1A functions is a role in DS mental deficits, as deduced from the essential function of the *Mnb* gene in *Drosophila* neurogenesis (Tejedor et al., 1995). In the worm *Caenorhabditis elegans*, the addition of an extra copy of its DYRK1A ortholog, mbk-1, to the genome causes behavioral abnormalities in chemotaxis and olfactory defects (Raich et al., 2003). The critical role of DYRK1A in DS mental retardation was further strengthened by analyzing a variety of genetically modified mice. Dyrk1A knockout mice show a general delay in fetal development and are embryonically lethal, indicating the vital and non-redundant biological functions of Dyrk1A (Fotaki et al., 2002). Transgenic mice that overexpress human DYRK1A or mouse Dyrk1A protein show learning and cognitive deficits similar to those seen in DS patients (Ahn et al., 2006; Altafaj et al., 2001; Smith et al., 1997).

Recently, several studies supporting the hypothesis that DYRK1A is involved in early-onset AD in the brains of DS have been reported. Chromosome 21 markers were scanned to assess genetic associations of late-onset AD patient samples. Among over 400 markers covering 70% of chromosome 21, 33 Mb, the DYRK1A gene showed the highest significance (Kimura *et al.*, 2007). When compared with normal control samples, samples from AD patients showed increased DYRK1A mRNA levels and immunoreactivity in the frontal cortex, entorhinal cortex, and hippocampus (Ferrer et al., 2005; Kimura et al., 2007), suggesting a potential role of DYRK1A in AD pathogenesis. Kimura *et al.* reported that Aβ treatment induced increased DYKR1A transcript, leading to the phosphorylation of Tau protein at Thr212. This demonstrates a potential role of DYRK1A in linking the gap between Aβ production and Tau hyperphosphorylation in AD (Kimura et al., 2007). Park *et al.* also reported the enhanced formation of Tau inclusion and Aβ in Dyrk1A-overexpressing immortalized hippocampal cells (Park et al., 2007).

The authors recently provided *in vivo* evidence of a physiological role of DYRK1A in APP processing and Tau hyperphosphorylation, suggesting a functional association between the overexpression of DYRK1A and early-onset AD in DS patients (Ryoo *et al.*, 2008; Ryoo *et al.*, 2007). It was demonstrated that DYRK1A phosphorylates APP at Thr668 as well as Tau at Ser202 and Ser404 *in vitro*. It was also found that the amounts of phospho-Thr668-APP, phospho-Ser202-Tau, phospho-Thr212-Tau, and phospho-Ser404-Tau are increased in the brains of transgenic mice that overexpress human DYRK1A. Furthermore, DYRK1A transgenic mouse brains were found to contain elevated amounts of Aβ. These results suggest that the overexpression of DYRK1A in DS regulates APP and Tau processing, thus constituting a potential underlying cause of the early-onset AD observed in DS individuals. On the basis of these findings, it is suggested that overexpression of DYRK1A in DS brains accelerates the development of AD pathogenesis through phosphorylation of APP and Tau, and that the extra copy of the *DYRK1A* gene plays a major role in the development of AD pathology for DS-affected individuals.

The finding that DYRK1A TG mice without an extra copy of the *APP* gene produce an increased amount of Aβ compared to control mice suggests that DYRK1A modulates APP processing through phosphorylation at Thr668 (Ryoo et al., 2008). The level of phospho-Thr668-APP was shown to be elevated in AD brains, and the mutation of Thr668 to Ala resulted in a decrease in Aβ production (Lee et al., 2003). APP phosphorylation at Thr668 may result in a conformational change that affects the interaction of APP with its binding partners (Ramelot et al., 2001). It was reported that Thr668 phosphorylation may enhance APP processing through β- and γ-secretase (Lee et al., 2003; Vingtdeux et al., 2005).

Phosphorylation of Tau at Ser202, Thr205 and Thr212 residues has been reported to enhance Tau polymerization and filament formation by increasing the level of stability or by slowing the dissociation rates (Braak et al., 1994; Rankin et al., 2005). It has been reported that Tau phosphorylation at Ser396/pSer404 is specifically increased in AD brain sections and that this phosphorylation enhances Tau polymerization and inhibits microtubule assembly (Abraha et al., 2000; Ding et al., 2006; Ryoo et al., 2007). Therefore, the increased Tau phosphorylation levels at Ser202, Thr212, and Ser404 seen in DYRK1A TG mice suggests an important role of DYRK1A in the formation of the Tau filaments that are necessary for the formation of NFTs (Ryoo et al., 2007).

As DYRK1A phosphorylates or interacts with a variety of proteins, the possibility remains that DYRK1A indirectly regulates APP and Tau phosphorylation *in vivo* through the activation of the protein kinase(s) and/or inhibition of the protein phosphatase(s) implicated in APP and Tau processing. DYRK1A also phosphorylates several proteins implicated in endocytosis, including Dynamin 1, Amphiphysin 1, and Synaptojanin 1, suggesting the role of DYRK1A in the abnormal endosome pathology present in AD and DS (Adayev et al., 2006; Chen-Hwang et al., 2002; Murakami et al., 2006).

ABCG1, BACE2, and RUNX1 genes.

Recently, Tansley *et al.* found another gene on chromosome 21 that encodes cholesterol transporter ABCG1, which may cause the early onset of AD in DS individuals (Tansley et al., 2007). They showed that ABCG1 influences the distribution and processing of APP and that it is present at high levels in the brains of DS individuals, suggesting that altered cholesterol metabolism and APP trafficking mediated by ABCG1 contributes to the accelerated onset of AD neuropathology in DS.

Aβ is the central component of senile plaques. It is generated from APP by means of cleavage by the β- and γ-secretases. BACE1 is the major β-secretase *in vivo*, and BACE2 is the homologue of BACE1. Given that the *BACE2* gene is located on chromosome 21, it was speculated that BACE2 plays a role in AD pathogenesis in DS. Although conflicting results have been reported (Motonaga et al., 2002; Myllykangas et al., 2005), recent data has shown that the expression level of BACE2 is unchanged despite the presence of an extra copy of the *BACE2* gene in DS, implying that BACE2 is not involved in the AD pathogenesis of DS patients (Sun et al., 2006).

From the screening of chromosome 21 markers, Kimura *et al.* reported that the RUNX1/CBFA2/AML1 gene shows a high odds ratio (OR = 23.3, $P<0.05$) in AD samples, implying an association of RUNX1 with late-onset AD. RUNX1 is a runt-related transcription factor that is widely expressed in multiple hematopoietic cells and is a critical regulator in hematopoiesis (Gao et al., 1991; Miyoshi et al., 1991; Speck et al., 1999). Haploinsufficiency of RUNX1 is known to cause a familial platelet defect with propensity to develop acute myelogenous leukemia (Song et al., 1999). Based on the function of RUNX1, it may be that the high incidence of leukemia seen in children with DS is due to the overexpression of RUNX1. Regarding AD, however, the potential mechanism of RUNX1 involved in the pathogenesis of AD remains unknown.

Conclusion and Summary

In this review, recent studies that enlighten the potential mechanism involved in the early AD onset in DS are summarized with a focus on the role of DYRK1A-mediated phosphorylation. A growing number of studies now show that DYRK1A is involved in early-onset AD in the brains of DS individuals through the regulation of Aβ production and Tau hyperphosphorylation by DYRK1A, either directly or indirectly. It is expected that the list of DYRK1A substrates will continue to grow and that an increasing amount of evidence will support roles for DYRK1A in the pathogenesis of AD. However, a clearer picture of the DYRK1A functions will depend on progress made by researchers who study AD pathogenesis. As therapeutic approaches aimed at reducing the levels of Aβ and abnormal

Tau phosphorylation may work for the treatment of AD, DYRK1A may possibly serve as a potential therapeutic target for the treatment of AD in both DS and AD patients.

ACKNOWLEDGMENTS

This work was supported by KOSEF (R01-2007-000-11910-0) funded by the Korean government (MOST). It was also supported by a Korea Research Foundation grant funded by the Korean Government (MOEHRD) (KRF-2007-331-E00031 to W-JS and KRF-2007-331-E00198 to S-HC) and by the IBST Grant 2007. The authors regret that they were not able to cite all relevant publications in this review.

REFERENCES

Abraha, A., Ghoshal, N., Gamblin, T.C., Cryns, V., Berry, R.W., Kuret, J., Binder, L.I. C-terminal inhibition of tau assembly in vitro and in Alzheimer's disease. *J. Cell. Sci.* **113**: 3737-3745, 2000.

Adayev, T., Chen-Hwang, M.C., Murakami, N., Wang, R., Hwang, Y.W. MNB/DYRK1A phosphorylation regulates the interactions of synaptojanin 1 with endocytic accessory proteins. *Biochem. Biophys. Res. Commun.* **351**: 1060-1065, 2006.

Ahn, K.J., Jeong, H.K., Choi, H.S., Ryoo, S.R., Kim, Y.J., Goo, J.S., Choi, S.Y., Han, J.S., Ha, I., Song, W.J. DYRK1A BAC transgenic mice show altered synaptic plasticity with learning and memory defects. *Neurobiol. Dis.* **22**: 463-472, 2006.

Altafaj, X., Dierssen, M., Baamonde, C., Marti, E., Visa, J., Guimera, J., Oset, M., Gonzalez, J.R., Florez, J., Fillat, C., Estivill, X. Neurodevelopmental delay, motor abnormalities and cognitive deficits in transgenic mice overexpressing Dyrk1A (minibrain), a murine model of Down's syndrome. *Hum. Mol. Genet.* **10**: 1915-1923, 2001.

Bielska, A.A., Zondlo, N.J. Hyperphosphorylation of tau induces local polyproline II helix. *Biochemistry* **45**: 5527-5537, 2006.

Braak, E., Braak, H., Mandelkow, E.M. A sequence of cytoskeleton changes related to the formation of neurofibrillary tangles and neuropil threads. *Acta Neuropathol. (Berl)* **87**: 554-567, 1994.

Burger, P.C., Vogel, F.S. The development of the pathologic changes of Alzheimer's disease and senile dementia in patients with Down's syndrome. *Am. J. Pathol.* **73**: 457-476, 1973.

Cataldo, A.M., Petanceska, S., Peterhoff, C.M., Terio, N.B., Epstein, C.J., Villar, A., Carlson, E.J., Staufenbiel, M., Nixon, R.A. App gene dosage modulates endosomal abnormalities of Alzheimer's disease in a segmental trisomy 16 mouse model of down syndrome. *J. Neurosci.* **23**: 6788-6792, 2003.

Cataldo, A.M., Petanceska, S., Terio, N.B., Peterhoff, C.M., Durham, R., Mercken, M., Mehta, P.D., Buxbaum, J., Haroutunian, V., Nixon, R.A. Abeta localization in abnormal endosomes: association with earliest Abeta elevations in AD and Down syndrome. *Neurobiol. Aging* **25**: 1263-1272, 2004.

Cataldo, A.M., Peterhoff, C.M., Troncoso, J.C., Gomez-Isla, T., Hyman, B.T., Nixon, R.A. Endocytic pathway abnormalities precede amyloid beta deposition in sporadic

Alzheimer's disease and Down syndrome: differential effects of APOE genotype and presenilin mutations. *Am. J. Pathol.* **157**: 277-286, 2000.

Chen-Hwang, M.C., Chen, H.R., Elzinga, M., Hwang, Y.W. Dynamin is a minibrain kinase/dual specificity Yak1-related kinase 1A substrate. *J. Biol. Chem.* **277**: 17597-17604, 2002.

De Strooper, B. Aph-1, Pen-2, and Nicastrin with Presenilin generate an active gamma-Secretase complex. *Neuron* **38**: 9-12, 2003.

Ding, H., Matthews, T.A., Johnson, G.V. Site-specific phosphorylation and caspase cleavage differentially impact tau-microtubule interactions and tau aggregation. *J. Biol. Chem.* **281**: 19107-19114, 2006.

Dowjat, W.K., Adayev, T., Kuchna, I., Nowicki, K., Palminiello, S., Hwang, Y.W., Wegiel, J. Trisomy-driven overexpression of DYRK1A kinase in the brain of subjects with Down syndrome. *Neurosci. Lett.* **413**: 77-81, 2007.

Ferrer, I., Barrachina, M., Puig, B., Martinez de Lagran, M., Marti, E., Avila, J., Dierssen, M. Constitutive Dyrk1A is abnormally expressed in Alzheimer disease, Down syndrome, Pick disease, and related transgenic models. *Neurobiol. Dis.* **20**: 392-400, 2005.

Fotaki, V., Dierssen, M., Alcantara, S., Martinez, S., Marti, E., Casas, C., Visa, J., Soriano, E., Estivill, X., Arbones, M.L. Dyrk1A haploinsufficiency affects viability and causes developmental delay and abnormal brain morphology in mice. *Mol. Cell. Biol.* **22**: 6636-6647, 2002.

Galceran, J., de Graaf, K., Tejedor, F.J., Becker, W. The MNB/DYRK1A protein kinase: genetic and biochemical properties. *J. Neural Transm. Suppl.* **67**: 139-148, 2003.

Gao, J., Erickson, P., Gardiner, K., Le Beau, M.M., Diaz, M.O., Patterson, D., Rowley, J.D., Drabkin, H.A. Isolation of a yeast artificial chromosome spanning the 8;21 translocation breakpoint t(8;21)(q22;q22.3) in acute myelogenous leukemia. *Proc. Natl. Acad. Sci. U S A* **88**: 4882-4886, 1991.

Grundke-Iqbal, I., Iqbal, K., Tung, Y.C., Quinlan, M., Wisniewski, H.M., Binder, L.I. Abnormal phosphorylation of the microtubule-associated protein tau (tau) in Alzheimer cytoskeletal pathology. *Proc. Natl. Acad. Sci. U S A* **83**: 4913-4917, 1986.

Guimera, J., Casas, C., Estivill, X., Pritchard, M. Human minibrain homologue (MNBH/DYRK1): characterization, alternative splicing, differential tissue expression, and overexpression in Down syndrome. *Genomics* **57**: 407-418, 1999.

Guimera, J., Casas, C., Pucharcos, C., Solans, A., Domenech, A., Planas, A.M., Ashley, J., Lovett, M., Estivill, X., Pritchard, M.A. A human homologue of Drosophila minibrain (MNB) is expressed in the neuronal regions affected in Down syndrome and maps to the critical region. *Hum. Mol. Genet.* **5**: 1305-1310, 1996.

Hammerle, B., Carnicero, A., Elizalde, C., Ceron, J., Martinez, S., Tejedor, F.J. Expression patterns and subcellular localization of the Down syndrome candidate protein MNB/DYRK1A suggest a role in late neuronal differentiation. *Eur. J. Neurosci.* **17**: 2277-2286, 2003.

Hanger, D.P., Brion, J.P., Gallo, J.M., Cairns, N.J., Luthert, P.J., Anderton, B.H. Tau in Alzheimer's disease and Down's syndrome is insoluble and abnormally phosphorylated. *Biochem. J.* **275 (Pt 1)**: 99-104, 1991.

Iwatsubo, T., Mann, D.M., Odaka, A., Suzuki, N., Ihara, Y. Amyloid beta protein (A beta) deposition: A beta 42(43) precedes A beta 40 in Down syndrome. *Ann. Neurol.* **37**: 294-299, 1995.

Jacobs, P.A., Baikie, A.G., Court Brown, W.M., Strong, J.A. The somatic chromosomes in mongolism. *Lancet* **1**: 710, 1959.

Johnson, G.V., Stoothoff, W.H. Tau phosphorylation in neuronal cell function and dysfunction. *J. Cell Sci.* **117**: 5721-5729, 2004.

Kentrup, H., Becker, W., Heukelbach, J., Wilmes, A., Schurmann, A., Huppertz, C., Kainulainen, H., Joost, H.G. Dyrk, a dual specificity protein kinase with unique structural features whose activity is dependent on tyrosine residues between subdomains VII and VIII. *J. Biol. Chem.* **271**: 3488-3495, 1996.

Kimura, R., Kamino, K., Yamamoto, M., Nuripa, A., Kida, T., Kazui, H., Hashimoto, R., Tanaka, T., Kudo, T., Yamagata, H., Tabara, Y., Miki, T., Akatsu, H., Kosaka, K., Funakoshi, E., Nishitomi, K., Sakaguchi, G., Kato, A., Hattori, H., Uema, T., Takeda, M. The DYRK1A gene, encoded in chromosome 21 Down syndrome critical region, bridges between beta-amyloid production and tau phosphorylation in Alzheimer disease. *Hum. Mol. Genet.* **16**: 15-23, 2007.

Kopke, E., Tung, Y.C., Shaikh, S., Alonso, A.C., Iqbal, K., Grundke-Iqbal, I. Microtubule-associated protein tau. Abnormal phosphorylation of a non-paired helical filament pool in Alzheimer disease. *J. Biol. Chem.* **268**: 24374-24384, 1993.

Korenberg, J.R., Chen, X.N., Schipper, R., Sun, Z., Gonsky, R., Gerwehr, S., Carpenter, N., Daumer, C., Dignan, P., Disteche, C., et al. Down syndrome phenotypes: the consequences of chromosomal imbalance. *Proc. Natl. Acad. Sci. U S A* **91**: 4997-5001, 1994.

Kosik, K.S. Alzheimer's disease: a cell biological perspective. *Science* **256**: 780-783, 1992.

Ksiezak-Reding, H., Liu, W.K., Yen, S.H. Phosphate analysis and dephosphorylation of modified tau associated with paired helical filaments. *Brain Res.* **597**: 209-219, 1992.

Lee, M.S., Kao, S.C., Lemere, C.A., Xia, W., Tseng, H.C., Zhou, Y., Neve, R., Ahlijanian, M.K., Tsai, L.H. APP processing is regulated by cytoplasmic phosphorylation. *J. Cell Biol.* **163**: 83-95, 2003.

Lee, M.S., Tsai, L.H. Cdk5: one of the links between senile plaques and neurofibrillary tangles? *J. Alzheimers Dis.* **5**: 127-137, 2003.

Lejeune, J., Gautier, M., Turpin, R. [Study of somatic chromosomes from 9 mongoloid children.]. *C R Hebd Seances Acad. Sci.* **248**: 1721-1722, 1959.

Lochhead, P.A., Sibbet, G., Morrice, N., Cleghon, V. Activation-Loop Autophosphorylation Is Mediated by a Novel Transitional Intermediate Form of DYRKs. *Cell* **121**: 925-936, 2005.

Mann, D.M. The pathological association between Down syndrome and Alzheimer disease. *Mech. Ageing Dev.* **43**: 99-136, 1988.

Marti, E., Altafaj, X., Dierssen, M., de la Luna, S., Fotaki, V., Alvarez, M., Perez-Riba, M., Ferrer, I., Estivill, X. Dyrk1A expression pattern supports specific roles of this kinase in the adult central nervous system. *Brain Res.* **964**: 250-263, 2003.

Masters, C.L., Simms, G., Weinman, N.A., Multhaup, G., McDonald, B.L., Beyreuther, K. Amyloid plaque core protein in Alzheimer disease and Down syndrome. *Proc. Natl. Acad. Sci. U S A* **82**: 4245-4249, 1985.

Miyoshi, H., Shimizu, K., Kozu, T., Maseki, N., Kaneko, Y., Ohki, M. t(8;21) breakpoints on chromosome 21 in acute myeloid leukemia are clustered within a limited region of a single gene, AML1. *Proc. Natl. Acad. Sci. U S A* **88**: 10431-10434, 1991.

Motonaga, K., Itoh, M., Becker, L.E., Goto, Y., Takashima, S. Elevated expression of beta-site amyloid precursor protein cleaving enzyme 2 in brains of patients with Down syndrome. *Neurosci. Lett.* **326**: 64-66, 2002.

Murakami, N., Xie, W., Lu, R.C., Chen-Hwang, M.C., Wieraszko, A., Hwang, Y.W. Phosphorylation of amphiphysin I by minibrain kinase/dual-specificity tyrosine phosphorylation-regulated kinase, a kinase implicated in Down syndrome. *J. Biol. Chem.* **281**: 23712-23724, 2006.

Myllykangas, L., Wavrant-De Vrieze, F., Polvikoski, T., Notkola, I.L., Sulkava, R., Niinisto, L., Edland, S.D., Arepalli, S., Adighibe, O., Compton, D., Hardy, J., Haltia, M., Tienari, P.J. Chromosome 21 BACE2 haplotype associates with Alzheimer's disease: a two-stage study. *J. Neurol. Sci.* **236**: 17-24, 2005.

Nixon, R.A. Endosome function and dysfunction in Alzheimer's disease and other neurodegenerative diseases. *Neurobiol. Aging.* **26**: 373-382, 2005.

Olson, M.I., Shaw, C.M. Presenile dementia and Alzheimer's disease in mongolism. *Brain* **92**: 147-156, 1969.

Park, J., Yang, E.J., Yoon, J.H., Chung, K.C. Dyrk1A overexpression in immortalized hippocampal cells produces the neuropathological features of Down syndrome. *Mol. Cell. Neurosci.* **36**: 270-279, 2007.

Patterson, D. The causes of Down syndrome. *Sci. Am.* **257**: 52-57, 60, 1987.

Prasher, V.P., Farrer, M.J., Kessling, A.M., Fisher, E.M., West, R.J., Barber, P.C., Butler, A.C. Molecular mapping of Alzheimer-type dementia in Down's syndrome. *Ann. Neurol.* **43**: 380-383, 1998.

Pulsifer, M.B. The neuropsychology of mental retardation. *J. Int. Neuropsychol. Soc.* **2**: 159-176, 1996.

Rahmani, Z., Lopes, C., Rachidi, M., Delabar, J.M. Expression of the mnb (dyrk) protein in adult and embryonic mouse tissues. *Biochem. Biophys. Res. Commun.* **253**: 514-518, 1998.

Raich, W.B., Moorman, C., Lacefield, C.O., Lehrer, J., Bartsch, D., Plasterk, R.H., Kandel, E.R., Hobert, O. Characterization of Caenorhabditis elegans homologs of the Down syndrome candidate gene DYRK1A. *Genetics* **163**: 571-580, 2003.

Ramelot, T.A., Nicholson, L.K. Phosphorylation-induced structural changes in the amyloid precursor protein cytoplasmic tail detected by NMR. *J. Mol. Biol.* **307**: 871-884, 2001.

Rankin, C.A., Sun, Q., Gamblin, T.C. Pseudo-phosphorylation of tau at Ser202 and Thr205 affects tau filament formation. *Brain Res. Mol. Brain Res.* **138**: 84-93, 2005.

Rovelet-Lecrux, A., Hannequin, D., Raux, G., Le Meur, N., Laquerriere, A., Vital, A., Dumanchin, C., Feuillette, S., Brice, A., Vercelletto, M., Dubas, F., Frebourg, T., Campion, D. APP locus duplication causes autosomal dominant early-onset Alzheimer disease with cerebral amyloid angiopathy. *Nat. Genet.* **38**: 24-26, 2006.

Ryoo, S.R., Cho, H.J., Lee, H.W., Jeong, H.K., Radnaabazar, C., Kim, Y.S., Kim, M.J., Son, M.Y., Seo, H., Chung, S.H., Song, W.J. Dual-specificity tyrosine(Y)-phosphorylation regulated kinase 1A-mediated phosphorylation of amyloid precursor protein: evidence for a functional link between Down syndrome and Alzheimer's disease. *J. Neurochem.* **104**: 1333-1344, 2008.

Ryoo, S.R., Jeong, H.K., Radnaabazar, C., Yoo, J.J., Cho, H.J., Lee, H.W., Kim, I.S., Cheon, Y.H., Ahn, Y.S., Chung, S.H., Song, W.J. DYRK1A-mediated hyperphosphorylation of

Tau. A functional link between Down syndrome and Alzheimer disease. *J. Biol. Chem.* **282**: 34850-34857, 2007.

Shindoh, N., Kudoh, J., Maeda, H., Yamaki, A., Minoshima, S., Shimizu, Y., Shimizu, N. Cloning of a human homolog of the Drosophila minibrain/rat Dyrk gene from "the Down syndrome critical region" of chromosome 21. *Biochem. Biophys. Res. Commun.* **225**: 92-99, 1996.

Smith, D.J., Stevens, M.E., Sudanagunta, S.P., Bronson, R.T., Makhinson, M., Watabe, A.M., O'Dell, T.J., Fung, J., Weier, H.U., Cheng, J.F., Rubin, E.M. Functional screening of 2 Mb of human chromosome 21q22.2 in transgenic mice implicates minibrain in learning defects associated with Down syndrome. *Nat. Genet.* **16**: 28-36, 1997.

Song, W.J., Sternberg, L.R., Kasten-Sportes, C., Keuren, M.L., Chung, S.H., Slack, A.C., Miller, D.E., Glover, T.W., Chiang, P.W., Lou, L., Kurnit, D.M. Isolation of human and murine homologues of the Drosophila minibrain gene: human homologue maps to 21q22.2 in the Down syndrome "critical region". *Genomics* **38**: 331-339, 1996.

Song, W.J., Sullivan, M.G., Legare, R.D., Hutchings, S., Tan, X., Kufrin, D., Ratajczak, J., Resende, I.C., Haworth, C., Hock, R., Loh, M., Felix, C., Roy, D.C., Busque, L., Kurnit, D., Willman, C., Gewirtz, A.M., Speck, N.A., Bushweller, J.H., Li, F.P., Gardiner, K., Poncz, M., Maris, J.M., Gilliland, D.G. Haploinsufficiency of CBFA2 causes familial thrombocytopenia with propensity to develop acute myelogenous leukaemia. *Nat. Genet.* **23**: 166-175, 1999.

Speck, N.A., Stacy, T., Wang, Q., North, T., Gu, T.L., Miller, J., Binder, M., Marin-Padilla, M. Core-binding factor: a central player in hematopoiesis and leukemia. *Cancer Res.* **59**: 1789s-1793s, 1999.

Sun, X., He, G., Song, W. BACE2, as a novel APP theta-secretase, is not responsible for the pathogenesis of Alzheimer's disease in Down syndrome. *FASEB J.* **20**: 1369-1376, 2006.

Tansley, G.H., Burgess, B.L., Bryan, M.T., Su, Y., Hirsch-Reinshagen, V., Pearce, J., Chan, J.Y., Wilkinson, A., Evans, J., Naus, K.E., McIsaac, S., Bromley, K., Song, W., Yang, H.C., Wang, N., DeMattos, R.B., Wellington, C.L. The cholesterol transporter ABCG1 modulates the subcellular distribution and proteolytic processing of beta-amyloid precursor protein. *J. Lipid Res.* **48**: 1022-1034, 2007.

Tate, B.A., Mathews, P.M. Targeting the role of the endosome in the pathophysiology of Alzheimer's disease: a strategy for treatment. *Sci. Aging Knowledge Environ.* **2006**: re2, 2006.

Tejedor, F., Zhu, X.R., Kaltenbach, E., Ackermann, A., Baumann, A., Canal, I., Heisenberg, M., Fischbach, K.F., Pongs, O. minibrain: a new protein kinase family involved in postembryonic neurogenesis in Drosophila. *Neuron* **14**: 287-301, 1995.

Vingtdeux, V., Hamdane, M., Gompel, M., Begard, S., Drobecq, H., Ghestem, A., Grosjean, M.E., Kostanjevecki, V., Grognet, P., Vanmechelen, E., Buee, L., Delacourte, A., Sergeant, N. Phosphorylation of amyloid precursor carboxy-terminal fragments enhances their processing by a gamma-secretase-dependent mechanism. *Neurobiol. Dis.* **20**: 625-637, 2005.

Wisniewski, K.E., Wisniewski, H.M., Wen, G.Y. Occurrence of neuropathological changes and dementia of Alzheimer's disease in Down's syndrome. *Ann. Neurol.* **17**: 278-282, 1985.

In:Cognitive Science Compendium Volume 2
Editors: Miao-Kun Sun

ISBN: 978-1-60456-775-5

Chapter 8

CHOLESTEROL, SECRETASE ACTIVITIES, AND Aβ PRODUCTION IN ALZHEIMER'S PATHOGENESIS

Wandong Zhang*

Neurobiology Program, Institute for Biological Sciences, National Research Council of Canada. Faculty of Medicine, University of Ottawa, Ottawa, Ontario, Canada K1A0R6

ABSTRACT

Alzheimer's disease (AD) is characterized by overproduction and deposition of β-amyloid peptides (Aβ) in the brain. The majority of AD patients are sporadic late-onset cases (LOAD), and the cause of Aβ overproduction and deposition in these patients is still largely unknown. Recent studies have linked altered cholesterol metabolism to Alzheimer's disease. Cholesterol homeostasis is found to be impaired in AD brain with significant cholesterol retention, increased β-and γ-secretase activities and Aβ production. Since α-secretase is located in phospholipids-rich domain and β-and γ-secretases reside in the cholesterol-rich rafts of plasma membrane, their activities are subjected to changes or levels of cellular cholesterol. Many studies have demonstrated that cholesterol is capable of regulating the activities of these secretases in cells. High levels of cellular cholesterol favor β-and γ-secretase pathway of APP processing for increased Aβ production; while cholesterol depletion or statins stimulate α-secretase activities, inhibit β-and γ-secretases and reduce Aβ production. Impaired cholesterol homeostasis in AD brain may result from altered activities or levels of key nuclear receptors (such as LXR, RXR, PPAR or TR) involving regulation of cholesterol metabolism. Modulation of these nuclear receptors with agonists in cultured cells or animals has shown to reduce cellular cholesterol, β-and γ-secretase activities and Aβ production. ApoE is involved in lipid/cholesterol metabolism and redistribution in the brain. *ApoE4* allele is found to be a major risk factor of AD. Thus, ApoE4 may also contribute to impaired cholesterol homeostasis in Alzheimer's brain. Further investigations are needed to target impaired cholesterol homeostasis in AD brain to

* Corresponding author: Dr. Wandong Zhang, Institute for Biological Sciences, National Research Council of Canada, 1200 Montreal Road, Building M54, Ottawa, Ontario, Canada K1A0R6, Tel: (613) 993-5988, Fax: (613) 941-4475, E-mail: Wandong.Zhang@nrc-cnrc.gc.ca

modulate cholesterol metabolism and β-and γ-secretase activities and to reduce Aβ production.

I. Introduction

Alzheimer's disease (AD) is a multi-facet neurodegenerative disease characterized by the overproduction and deposition of β-amyloid peptides (Aβ) and the formation of intracellular neurofibrillary tangles (NFT) in the brain (1 - 4). Aβ peptides are aggregated and deposited as amyloid plaques (neuritic or senile plaques) in brain cortex (1 - 4). Recent studies suggest that soluble oligomers of Aβ peptides are the toxic form and affect the synaptic functions (1 - 4). Both Aβ oligomers and plaques are believed to contribute to neurodegenerative pathology and cognitive deficits in AD patients (1 - 4). The pathogenic mechanism of Alzheimer's disease is still debatable; however, a number of epidemiological, genetic and experimental studies support the β-amyloid hypothesis for the disease. Familial AD (FAD) patients carry inherited mutations in genes encoding β-amyloid precursor protein (APP) (e.g., Swedish double mutation) or γ-secretase component presenilins (PSEN1, PSEN2) (1 - 4). A large amount of Aβ peptides is produced and deposited in the brains of FAD, leading to early onset of AD. Down's syndrome with three copies of APP genes on tri-chromosome 21 results in overproduction and deposition of Aβ peptides in the brain, which is thought to be responsible for mental retardation and gradual cognitive decline in these individuals. A number of transgenic mouse models with mutant hAPP or/and PSEN1 have been established, and overproduction and deposition of Aβ peptides are found in the brain of these transgenic mice, which results in Alzheimer's pathology and memory deficits in the animals. Clinical trials and animal experiments with active beta-amyloid immunization or anti-amyloid antibodies show reversible AD pathology and improved AD symptoms/cognitive performance. All of these studies support the amyloid hypothesis. However, over 95% of the AD patients are sporadic late-onset cases (LOAD) (5, 6). It is unknown what causes Aβ overproduction in LOAD. Aβ peptide is a product derived from proteolytic cleavage of APP by secretases. Altered proteolytic processing of APP may lead to incrased Aβ production in LOAD.

II. Secretases and Processing Pathways for APP

Aβ peptide is derived from sequential β-and γ-secretase cleavage of APP anchored in plasma membrane (3, 7). During and after the trafficking of APP through secretory pathway, APP undergoes a variety of proteolytic cleavage to release secreted derivatives. Three secretases, α, β, and γ, are involved in APP processing (3, 7). α-Secretase-mediated cleavage occurs in the Aβ domain of APP and generates a soluble N-terminal fragment (APPs-α) and a membrane-bound carboxyl-terminal fragment-α (CTF-α or C83). C83 is then cleaved by γ-secretase to generate a 3 kDa fragment (p3) and an APP intracellular domain (AICD). Alternatively, cleavage of APP by β-secretase generates a soluble N-terminal fragment (APPs-β) and a membrane-bound carboxyl-terminal fragment (CTF-β or C99) (3, 7). C99 bears the total Aβ domain and is subsequently cleaved by γ-secretase, yielding Aβ and AICD.

Therefore, sequential cleavage of APP by α-and γ-secretases does not generate Aβ peptide; while sequential cleavage of APP by β-and γ-secretases, e.g., β-and γ-secretase pathway of APP processing, releases $A\beta_{1-40}$ or $A\beta_{1-42}$ (3, 7). β-Secretase cleavage site is located 28 amino acids away from the extracellular face of the membrane; while the γ-secretase cleavage site is located within the lipid bilayer (1 - 3). γ-Secretase cleavage occurs at multiple sites. The dominant cleavage occurs at 12 amino acids from the C-terminus, resulting in $A\beta_{1-40}$ peptide and the less common cleavage occurs at 14 amino acids from the C-terminus producing $A\beta_{1-42}$ peptide. $A\beta_{1-40}$ is primarily found in vascular deposits; while $A\beta_{1-42}$ is mainly found in neuritic plaques (1 - 3). Aβ peptides are amphipathic, consisting of 28 hydrophilic and 12-14 hydrophobic amino acids. The production and release of Aβ is a normal metabolic event in cells and the peptides can be detected in cerebral spinal fluid (CSF) and plasma in health subjects throughout life (1 - 3). The physiological role of Aβ is not clear. A recent study indicates that the production of endogenous Aβ is critical for the viability of central neurons (8). Aβ is also considered as a regulator of lipid homeostasis (9). It is unknown what factor(s) determines the switch between the two pathways of APP processing in the brain. Since both β-and γ-secretase complexes reside in cholesterol-rich lipid rafts of plasma membrane and α-secretase is located in phospholipid-rich domains of the membrane, their activities may be affected by altered levels of cholesterol and the ratio of cholesterol to phospholipids in plasma membrane (5, 6, 10, 11). It is generally believed that a membrane domain rich in cholesterol favors β-and γ-secretase pathway of APP processing for Aβ production (5, 6, 10, 11). Kaether and Haass (12) suggests that there is a dose-dependent effect of cholesterol reduction on Aβ formation, e.g., moderate reduction of membrane cholesterol leads to increased Aβ formation; whereas strong reduction of membrane cholesterol results in a significant drop in Aβ generation (12, 13). Thus, secretase activities and Aβ formation may be subjected to the changes of cellular cholesterol.

III. Cholesterol and Alzheimer's Disease

1. Cholesterol metabolism and homeostasis in the brain

Cholesterol is an important component of cell membrane. There are two sources of cholesterol in human body, absorption from food ingested in gut and *de novo* synthesis in cells and tissues. It is estimated that 1g of cholesterol is synthesized *de novo* per day, and 0.3g of cholesterol is absorbed from the gut per day (14 - 16). Cholesterol cannot be degraded, but rather converted to steroid hormones, bile acids, oxidized into oxysterols or redistributed among tissues via cellular cholesterol efflux to lipoproteins. Transport of fatty acids and cholesterol is mainly mediated by four classes of lipoproteins in human body, including chylomicron, very low-density lipopoprotein (VLDL), LDL, and high-density lipopoprotein (HDL). Lipoproteins are made up of a hydrophilic phospholipid monolayer exterior, specific proteins (called apolipoproteins) associated with the phospholipid surface, and an apolar lipid core. Among those lipoproteins, HDL is mainly involved in transport of cholesterol from extraheptic tissues to the liver. A number of apolipoproteins are associated with HDL, such as apo-A (AI, AII), apo-B, apo-C (CI, CII, CIII), apo-D, apo-E (E2, E3, E4), and apo-J, and serve as lipid receptors on HDL (5). Their expression and interactions determine lipid and cholesterol transport/redistribution in tissues. Apolipoproteins are synthesized by many cell types, including brain cells (astroglia is a main source and neurons express low levels of apolipoproteins) (5). ApoE is known as a major apolipoprotein in the brain and plays a key role in brain cholesterol redistribution. ApoE4 is associated with a high risk of AD. A recent report showed that apoE modulates γ-secretase cleavage of APP (17), in which apoE reduces γ-secretase cleavage of APP and secreted Aβ with stronger effects on $A\beta_{1-40}$, and suggested that this apoE modulation of Aβ production and APP signaling is a potential mechanism affecting Alzheimer disease risk (17).

Brain is especially rich in cholesterol. Compared to other tissues, brain is unique in terms of cholesterol metabolism and requirements. Brain contains approximately 25% of the total amount of cholesterol in the human body, although it accounts for only 2% of the body mass (14 - 16). Brain tissue is isolated from blood and other tissues by the blood-brain barrier (BBB) and its cholesterol (95%) is almost entirely synthesized *de novo* (14 - 16, 18). Over 99% of brain cholesterol is unesterified (14 - 16, 18). Brain lipids are primarily present in myelin sheaths (e.g., oligodendroglia) and plasma membranes of neurons and astrocytes. About 70% of cholesterol is present in myelin. The white matter of the brain is mainly composed of myelin and therefore richer in cholesterol (14 - 16). The cortex of the brain where the Aβ peptides are aggregated and deposited as amyloid plaques (neuritic or senile plaques) is mainly composed of neurons and astrocytes.

Cholesterol in the brain is actively turned over among neuronal and glial cells and is essential for neuronal synaptic plasticity and other important functions (14 - 16). Cholesterol turnover across the brain among astrocytes and neurons is mainly mediated by apolipoproteins (such as apoE) and cholesterol efflux transporter ABCA1 (5, 14 - 16). A small portion of brain cholesterol is oxidized by the enzyme, cholesterol 24-hydroxylase (CYP46A1), into 24S-hydroxylcholesterol which is then excreted from the brain through BBB (6-7mg/day) (15, 19). Some cholesterol may be excreted from the brain via ABCA1 expressed at the BBB, but this pathway is not well defined (15). ApoE-bound cholesterol can be excreted from the brain via CSF route (1-2mg/day) (15). The expression of the genes involving cholesterol metabolism and redistribution are mostly regulated by nuclear receptors LXR or/and RXR, including apolipoproteins (such as apoE etc), cholesterol efflux transporters (ABCA1, ABCG1, ABCG4), cholesterol 24-hydroxylase (CYP46A1) and others (15, 19 - 28). Nuclear receptor PPRA-γ agonist was shown to reduce cholesterol synthesis in cultured cells (29).

2. Regulation of cholesterol metabolism and homeostasis

A number of studies have shown that nuclear receptors play important roles in the regulation of cholesterol and lipid metabolism and in maintaining cholesterol homeostasis in cells and tissues (23 - 25, 30). Nuclear receptors are ligand-inducible transcription factors comprised of a superfamily with six subfamilies and 26 groups (25, 30). Ligand-activated nuclear receptors can form homodimers or heterodimiers with each other and then bind to and activate transcription of target genes (25, 30). The nuclear receptors involved in the regulation of cholesterol and lipid metabolism include liver X receptor (LXR), retinoid X receptor

(RXR), retinoic acid receptors (RAR), peroxisome proliferation-activated receptors (PPAR) (28), thyroid hormone receptor (TR), androgen receptors (AR), estrogen receptors (ER), progesterone receptors (PR), glucocorticoid receptors (GR), vitamin D receptors (VDR), and sterol receptor-associated proteins. Studies using macrophages/monocytes and cultured cells have demonstrated that LXR, RXR, and PPAR are capable of regulating cellular cholesterol metabolism and homeostasis (25, 31 - 36), and especially activation of LXR, RXR and PPAR is shown to improve cholesterol homeostasis and inhibit the development of atherosclerosis in animal experiments (31 - 35). Several recent studies have found that activation of TR has beneficial effects, including reduced serum cholesterol and body weight loss (37 - 39). There are two isoforms of TR, e.g., TRα1 and TRβ1. TRβ1 is more involved in lipid and cholesterol metabolism. Selective TR agonists have been designed for testing to reduce cholesterol and weight without adverse thyroid hormone action (36 - 39). These nuclear receptors regulate the expression of many genes involved in lipid and cholesterol metabolism. However, little is known about these receptors and their roles in Alzheimer's pathogenesis.

3. Cholesterol in AD Pathogenesis

Recent studies have linked altered cholesterol metabolism to Alzheimer's disease. It is suggested that altered cholesterol metabolism is involved in three major pathological processes described in AD, e.g., Aβ production and aggregation (formation of Aβ oligmers), and tau protein phosphorylation (40, 41). Cholesterol is known to interact with Aβ in a reciprocal manner, and cellular levels of cholesterol modulate Aβ generation (5, 6, 10 - 12). High cellular cholesterol load promotes Aβ formation, while decreased Aβ production is observed when *de novo* cholesterol synthesis is suppressed (42). Aβ alters cholesterol dynamics in neurons, leading to tauopathy (40, 41). A number of epidemiological, genetic and experimental studies have described the following observations.

(1) Hypercholesterolemia is an early risk factor for the development of Alzheimer amyloid pathology, especially high midlife serum total cholesterol is a risk factor of dementia/AD (18, 43 - 45). However, high total cholesterol levels in late life may not be a risk factor of dementia (46). Hypercholesterolemia accelerated the AD pathology and cholesterol-lowering drugs reduced Aβ pathology in a transgenic animal model (47, 48).
(2) Decreased prevalence of AD is associated with the use of cholesterol-lowering drug statins that inhibit 3-hydroxy-3-methylglutaryl coenzyme A reductase (HMG-CoA reductase inhibitors) (41, 49 - 51). The polymorphism of HMG-CoA reductase gene is associated with Alzheimer's risk and cognitive deterioration (52). Treatment with statins reduces serum Aβ peptides *in vivo* and the release of Aβ from cultured cells (53 - 56). Lovastatin was shown to inhibit beta-cleavage of APP (57). Some studies showed that the use of statin did not improve cognitive performance of AD patients and suggested that long-term study is necessary (56). Simvastatin enhances learning and memory in transgenic mice but is independent of amyloid load in the brain (58).
(3) Long-term dietary cholesterol increased cholesterol content in neurons and was associated with an increased level of BACE1 in the brain of rabbits (59).

(4) Aβ can bind HDL particles (including apoAI, apoA-II, apoE, and apoJ) and the binding inhibits Aβ aggregation (5, 11, 60, 61). Aβ is a regulator of lipid homeostasis (9). Cholesterol increases Aβ production, but Aβ production results in decreased cholesterol synthesis (9).

(5) Aβ formation in cells is inhibited by cholesterol depletion (62), and Aβ increases hippocampal lipid synthesis (63).

(6) Cholesterol is essential for synaptic plasticity and neuronal degeneration (64). Increased cholesterol efflux impairs short-and long-term synaptic plasticity and causes hippocampal neural degeneration (63).

(7) The ratio of cholesterol to phospholipids in the membrane lipid bilayer determines the insertion of $A\beta_{1-40}$ and fibril formation (65). Higher ratio appears to inhibit Aβ fibril formation.

(8) ABCA1 is a cholesterol transporter mediating cellular cholesterol efflux from cells to HDL (5, 20 - 22). Several studies show that ABCA1 affects APP processing and Aβ secretion via cellular cholesterol efflux to HDL in neuroblastoma N2a or brain cells (20 - 22). Overexpression of ABCA1 reduces Aβ deposits in PDAPP transgenic mice (66); while deletion of ABCA1 increases Aβ deposition in these mice (67). Genetic variability of *ABCA1* modulates cholesterol levels in CSF by interfering with cholesterol homeostasis of central nervous system (CNS) and affects the age of AD onset (68). *ABCA1* gene polymorphisms also affect AD risk and are related to Aβ metabolism in the brain (69, 70). Another cholesterol transporter ABCG1 is found to modulate subcellular distribution and proteolytic processing of APP (71), and its polymorphism is associated with the risk of sporadic AD (72).

All of the evidence described above suggests that cholesterol is an important factor in the pathogenesis and development of late-onset Alzheimer's disease.

4. Cholesterol regulates secretase activities and Aβ production

Since α-secretase is located in phospholipid-rich domain of plasma membrane and both β-and γ-secretases reside in cholesterol-rich lipid rafts of plasma membrane (10 - 13, 73), it is believed that altered levels of cholesterol or/and the ratio of cholesterol to phospholipids in plasma membrane would affect the activities of these secretases and therefore determine the pathways for APP processing and Aβ genesis. Wolozin hypothesized that higher level of cellular cholesterol inhibits α-secretase and promotes β-and γ-secretase activities for Aβ production (10, 11); while Kaether and Haas proposed that a moderate low-level of membrane cholesterol increases Aβ production and extremely poor cholesterol environment inhibits Aβ production (12). Bodovitz and Klein showed that cholesterol inhibits α-secretase-mediated APP processing and reduces APPs-α production (74). Cholesterol depletion stimulates APPs-α production and reduces Aβ release from cells (75). Cholesterol-lowing drugs atorvastatin and lovastatin stimulate α-secretase-mediated APP processing with increased APPs-α release from treated cells (75, 76). Interestingly, a cholesterol derivative 24S-hydroxycholesterol which is reduced in AD brain stimulates α-secretase activity and increases APPs-α production (77). These studies indicate that lower cholesterol level favors α-secretase activity for reduced Aβ production. However, experimental studies yielded contradictory findings about the roles of cholesterol in the regulation of β-and γ-secretase activities for Aβ production. Simons and colleagues reported that cholesterol depletion inhibited Aβ production in hippocampal neurons (62). Cordy and colleagues showed that the GPI-anchored β-secretase was located in lipid rafts and Aβ secretion was reduced when lipid rafts were disrupted by depleting cellular cholesterol (78). In another recent study, cloned β-secretase was expressed in baculovirus system, purified and reconstituted *in vitro* with liposomes (79). The study identified 3 groups of lipids that stimulate proteolytic activity of BACE in reconstituted system, e.g., neutral glycosphingolipids (cerebrosides), anionic glycerophospholipids, and cholesterol (79). Wahrle and colleagues found that γ-secretase activity was located in cholesterol-rich membrane microdomains and was cholesterol-dependent (80). Depletion of membrane cholesterol completely inhibited γ-secretase activity and cholesterol replacement restored its activity (80). In contrary, Abad-Rodriguez and colleagues reported that membrane cholesterol loss in hippocampal neurons enhanced Aβ generation (13). Fukumoto and colleagues reported that stimulation of cholesterol efflux transporter ABCA1 (e.g., reduced cellular cholesterol) significantly increased $A\beta_{1\text{-}40}$ release from neuronal cells (20); while other studies showed that induction of ABCA1 expression reduced Aβ peptides in cells or in APP23 mice (21, 22, 81). These studies provide the evidence that cholesterol affects the activities of β- and γ-secretases or/and Aβ production in cultured cells or reconstituted *in vitro* systems. However, it is largely unknown whether cholesterol metabolism and its regulation are altered in AD brain and whether the activities of both β- and γ-secretases in human brain tissues can be regulated by cholesterol.

We have analyzed the levels of cholesterol in brain samples (occipital cortices) from age-matched non-demented controls (ND) and AD brain samples (occipital cortices) with pathology of cerebral amyloid angiopathy (AD/CAA). We have found that AD/CAA brains had significantly higher levels of cholesterol (19.43%) than ND by using fluorimetric enzymatic cholesterol assay (*t* test, $p = 0.0104$) (82). Over 99% of total brain cholesterol was free cholesterol (82). Chemical method analysis of total brain cholesterol confirmed that AD/CAA brains had significant higher level of cholesterol (43.1%, *t* test $p < 0.01$) than ND brain samples (82). Furthermore, we have also measured the levels of cholesterol in human AD brains without CAA pathology (occipital cortices) and found that level of cholesterol was also significantly higher in AD brains (27.15%↑, $p < 0.05$) than that of ND brains (82). This provides direct evidence that altered cholesterol metabolism does exist in AD brains. No significant gliosis was observed in AD brain samples, suggesting that higher level of cholesterol in these samples was not a result of gliosis. Our *in vitro* activity assays using human brain lysates show that both β-and γ-secretase activities were significantly higher in AD/CAA (22.2% and 39.8%) than in ND brain ($p < 0.01$), respectively (82). Western blot shows that the levels of BACE were similar in both ND and AD samples. To determine whether cholesterol would regulate/modulate the activities of these two secretases in human brain tissues, different concentrations of exogenous cholesterol were added to the *in vitro* assays using AD or ND brain lysates. The activities of both β- and γ-secretases were strongly stimulated by 80μM cholesterol in ND and by 40μM cholesterol in AD brain samples, respectively ($p < 0.01$) (82). Cholesterol at 20μM concentration inhibited γ-secretase activity in AD brain (82). These results demonstrate that cholesterol at optimal concentrations can stimulate the activities of β- and γ-secretases in human brain tissues and that β- and γ-secretases in AD brain are more sensitive to cholesterol. Our work suggests that cholesterol retention may be responsible for high β- and γ-secretase activities and Aβ production in AD brain.

To further validate the above findings from human brain samples, we used neuroblastoma 2a (N2a) cells-transfected with a human APP gene as an *in vitro* model system. Different concentrations of exogenous cholesterol (10, 20, 40, and 80 μM) were added to the cell cultures. The cells and media were harvested at 1, 2, 4 and 6 hr for cholesterol assay, β- and γ-secretase activity assays, and $A\beta_{1-40}$ ELISA, respectively (82). As we observed in cholesterol efflux assay using [^{3}H]-cholesterol (82), approximately 30-40% of cholesterol in culture media can be incorporated in cultured cells. The assays show that cholesterol levels in the cells with exogenous cholesterol in media were significantly increased compared to those of controls without cholesterol in media (One-way ANOVA $p < 0.05$) (82). The activities of β- and γ-secretases in cholesterol-incubated cells were significantly higher than those of controls (One-way ANOVA $p < 0.05$), especially at earlier time points (1, 2 and 4 h). Consequently, the secretion of $A\beta_{1-40}$ peptides into media from cholesterol-incubated cells was significantly increased in comparison with controls (One-way ANOVA $p < 0.0005$) (82). Furthermore, we prepared lysates from regular N2a cells and mixed lysates with different concentrations of cholesterol (20, 40, and 80 μM, respectively). The *in vitro* assays show that both β- and γ-secretase activities were strongly stimulated. It appears that 80 μM cholesterol was stronger than 20 and 40 μM cholesterol in stimulating β-secretase activity in N2a cell lysates; while 20 and 40 μM cholesterol was stronger than 80 μM cholesterol ($p < 0.05$) in stimulating γ-secretase activity in the lysates, suggesting that γ-secretase is more sensitive to cholesterol than β-secretase in N2a cells. N2a cells were treated with cholesterol depletion agent cyclodextran (CDT, 2.5 mM) for 2 h and the media were replaced with fresh media. Cells and media were harvested at 4, 8, 12 and 24h. Cholesterol was strongly decreased at 4 hr. β-Secretase activity and $A\beta_{1-40}$ secretion were significantly reduced, too. Cholesterol level, β-secretase activity and Aβ secretion were returned or above control levels at 8, 12 and 24 hr (82). These i*n vitro* experiments show that high cholesterol can stimulate both β- and γ-secretase activities in human brain tissues and cultured cells for increased production of Aβ peptides.

5. Altered nuclear receptors and impaired cholesterol homeostasis in Alzheimer's brains

Several studies have suggested that nuclear receptors LXR or RXR affect Aβ production in cultured neural cells or animals but the studies yield contrary results. One study showed that LXR agonists (TO901317) increased Aβ secretion from treated neuronal cells (20). Others showed that treatments with LXR agonists (TO901317, 22-R-hydroxylcholesterol) or RXR agonist (9-cis-retinoic acid) decreased Aβ production in cultured neuronal cells or in a mouse AD model (21, 22). There is no evidence whether the levels or activities of these nuclear receptors are altered in AD brain. We have analyzed and compared the activities or levels of these transcription factors in ND versus AD brain samples. Our profiling shows that the activities of PPAR and TR were decreased over two-fold and the activity of RXR was increased over two-fold in AD as compared to ND. The expression of LXR-β was significantly decreased in AD samples compared to ND samples ($p<0.01$). These results suggest that the altered levels or activities of these nuclear receptors (e.g., decreased PPAR, TR and LXR, and increased RXR) may be involved in the development of cholesterol retention in AD brain.

LXR and RXR are the main nuclear receptors that regulate the metabolism of lipids and cholesterol and maintain a balanced, fine-tuned regulation of cholesterol in cells (23 - 28). They function as homo-or heterodimers. A number of studies have shown that LXR or/and RXR agonists up-regulate the expression of lipid/cholesterol metabolism-related genes and cholesterol transporters (such as genes encoding apolipoproteins and ABC cholesterol transporters) and therefore reduce cellular cholesterol (20 - 22). One study showed that LXR agonists increase Aβ production in neural cells (20). Several others reported that LXR agonists reduce Aβ production in cells and AD mice (21, 22, 83). To clarify these contradictory findings and assess our findings on altered nuclear receptors in AD brain, we used cultured cells as *in vitro* models to test whether LXR or/and RXR would affect the levels of cellular cholesterol, β- and γ-secretase activities and Aβ production. N2a and K269 cells transfected with a human APP gene were cultured in the presence or absence of LXR or/and RXR agonists. Cholesterol efflux assay using K269 cells shows that 42.06% more [^{3}H]-labeled-cholesterol in cells treated with LXR agonist was transported to HDL compared to controls ($p < 0.01$) (82). Treatments of N2a cells with the agonists also reduced cellular cholesterol in N2a cells. *In vitro* assays show that both β- and γ-secretase activities were strongly decreased in N2a cells in the presence of the agonists. Similarly, treatment of N2a cells with a PPAR-γ agonist (troglitazone) significantly reduced both β- and γ-secretase activities (82). These results show that activation of nuclear receptors LXR, RXR or PPAR can reduce cellular cholesterol and therefore inhibit the activities of β- and γ-secretases and suggest that altered nuclear receptors may be responsible for cholesterol retention and impaired cholesterol homeostasis in AD brains. Restoration of cholesterol homeostasis will decrease β- and γ-secretase activities for less Aβ production. These findings suggest that the nuclear receptors may be served as possible therapeutic targets for modulation of cholesterol homeostasis, β- and γ-secretase activities and Aβ production in AD brain.

6. ApoE and Alzheimer's disease

Another important factor that may contribute to altered cholesterol metabolism/homeostasis in AD brain is apolipoprotein E (apoE). ApoE is a multifunctional protein with three common isoforms (apoE2, apoE3, and apoE4) and play an important part in lipid/cholesterol metabolism/homeostasis and neurobiology (83). ApoE is a major apolipoprotein in the brain. A number of epidemiological studies have suggested that apoE4 is associated with a high risk of AD (83) and also associated with rapid cognitive decline in elderly adults and may predispose to dementia (84). In addition to a number of other functions, ApoE is mainly involved in lipid/cholesterol redistribution in the brain and is a component of high-density lipoproteins (HDL) found in CSF where apoE accounts for the majority of lipoproteins (15, 83). Some studies suggest that apoE4 is associated with deficit in cholesterol transport (85). The structure of apoE4 molecule is bulky and may not be efficient in mediating cholesterol transport and redistribution compared to other apoE isoforms and lipoproteins. ApoE4 was found to disrupt sterol and sphingolipid metabolism in Alzheimer's but not normal brains (85). A recent study showed that apoE4 enhances Aβ production in cultured neuronal cells (86). Our genotyping reveals that over one-half of our AD patients examined had an apoE4 allele; while none of the ND cases had apoE4. This suggests the possible role of apoE4 in altered cholesterol metabolism for cholesterol retention and increased Aβ production in AD brain.

IV. Summary

Altered cholesterol metabolism and homeostasis is linked to Alzheimer's disease. Cholesterol retention is found in Alzheimer's brain with increased β- and γ-secretase activities and Aβ production. A number of studies have demonstrated that cholesterol is capable of regulating the activities of α-, β- and γ-secretases in cells. High level of cellular cholesterol inhibits α-secretase activity but increases β- and γ-secretase activities and Aβ production. In contrary, lower cholesterol stimulates α-secretase activity, inhibits β- and γ-secretase activities and reduces Aβ production in cells. Impaired cholesterol homeostasis may result from altered activities or levels of several key nuclear receptors in AD brain. Stimulation of these nuclear receptors with agonists reduces cellular cholesterol, β- and γ-secretase activities or Aβ production in cultured cells or animals. Since ApoE4 allele is a major risk factor of AD, apoE4 may also contribute to impaired cholesterol metabolism and homeostasis in AD brain. Targeting of altered cholesterol metabolism/homeostasis in Alzheimer's brain may lead to a potential new therapeutic strategy to modulate level of cholesterol, secretase activities and Aβ production.

Acknowledgments

The research in author's laboratory is supported by funding from the National Research Council of Canada, Heart & Stroke Foundation of Canada, Canadian Institute of Health Research, Alzheimer Society of Canada, Pfizer, and ApoPharma Inc.

References

[1] Selkoe, D.J. Alzheimer's disease: Genes, proteins, and therapy. *Physiol. Rev.* **81**: 741-766, 2001.
[2] Selkoe, D.J., Schenk, D. Alzheimer's disease: molecular understanding predicts amyloid-based therapeutics. *Annu. Rev. Pharmacol. Toxicol.* **43**: 545-584, 2002.
[3] Selkoe, D.J. Alzheimer disease: mechanistic understanding predicts novel therapies. *Ann Intern Med.* **140**:627-638, 2004.
[4] Hardy, J. A hundred years of Alzheimer's disease research. *Neuron.* **52**:3-13, 2006.
[5] Wellington, C.L. Cholesterol at the crossroads: Alzheimer's disease and lipid metabolism. *Clin. Genet.* **66**: 1-16, 2004.

[6] Puglielli, L., Tanzi, R.E., Kovacs, D.M. Alzheimer's disease: the cholesterol connection. *Nat Neurosci.* **6**:345-351, 2005.

[7] Walter, J., Kaether, C., Steiner, H., Haass, C. The cell biology of Alzheimer's disease: uncovering the secrets of secretases. *Curr Opin Neurobiol.* **11**:585-590, 2001.

[8] Plant, L.D., Boyle, J.P., Smith, I.F., Peers, C., Pearson, H.A. The production of amyloid beta peptide is a critical requirement for the viability of central neurons. *J Neurosci.* **23**: 5531-5535, 2003.

[9] Grimm, M.O., Grimm, H.S., Hartmann, T. Amyloid beta as a regulator of lipid homeostasis. *Trends Mol Med.* **13**:337-344, 2007.

[10] Wolozin, B. A fluid connection: Cholesterol and Aβ. *Proc Natl Acad Sci USA* **98**: 5371-5373, 2001.

[11] Wolozin, B. Cholesterol and the biology of Alzheimer's disease. *Neuron* **41**:7-10, 2004.

[12] Kaether, C., Haass, C. A lipid boundary separates APP and secretases and limits amyloid beta-peptide generation. *J Cell Biol.* **167**:809-812, 2004.

[13] Abad-Rodriguez, J., Ledesma, M.D., Craessaerts, K., Perga, S., Medina, M., Delacourte, A., Dingwall, C., De Strooper, B., Dotti, C.G. Neuronal membrane cholesterol loss enhances amyloid peptide generation. *J Cell Biol.* **167**:953-60, 2004.

[14] Dietschy, J.M., Turley, S.D. Cholesterol metabolism in the brain. *Curr Opin Lipidol.* **12**: 105-112, 2001.

[15] Bjorkhem, I., Meaney, S. Brain cholesterol: Long secret life behind a barrier. *Arterioscler Thromb Vasc Biol.* **24**: 806-815, 2004.

[16] Pfrieger, F.W. Outsourcing in the brain: do neurons depend on cholesterol delivery by astrocytes? *Bioessays* **25**:72-78, 2003.

[17] Irizarry, M.C,, Deng, A., Lleo, A., Berezovska, O., Von Arnim, C.A., Martin-Rehrmann, M., Manelli, A., LaDu, M.J., Hyman, B.T., Rebeck, G.W. Apolipoprotein E modulates gamma-secretase cleavage of the amyloid precursor protein. *J Neurochem.* **90**:1132-1143, 2004.

[18] Hartmann, T., Kuchenbecker, J., Grimm, M.O. Alzheimer's disease: the lipid connection. *J Neurochem.* **103** (Suppl 1):159-170, 2007.

[19] Brown ,J 3rd., Theisler, C., Silberman, S., Magnuson, D., Gottardi-Littell, N., Lee, J.M., Yager, D., Crowley, J., Sambamurti, K., Rahman, M.M., Reiss, A.B., Eckman, C.B., Wolozin, B. Differential expression of cholesterol hydroxylases in Alzheimer's disease. *J Biol Chem.* **279**:34674-34681, 2004.

[20] Fukumoto, H., Deng, A., Irizarry, M.C., Fitzgerald, M.L., Rebeck, G.W. Induction of the cholesterol transporter ABCA1 in central nervous system cells by liver X receptor agonists increases secreted Abeta levels. *J Biol Chem.* **277**:48508-48513, 2002.

[21] Koldamova, R.P., Lefterov, I.M., Ikonomovic, M.D., Skoko, J., Lefterov, P.I., Isanski, B.A., DeKosky, S.T., Lazo, J.S. 22R-hydroxycholesterol and 9-cis-retinoic acid induce ATP-binding cassette transporter A1 expression and cholesterol efflux in brain cells and decrease amyloid beta secretion. *J Biol Chem.* **278**:13244-13256, 2003.

[22] Sun, Y., Yao, J., Kim, T.W., Tall, A.R. Expression of liver X receptor target genes decreases cellular amyloid beta peptide secretion. *J Biol Chem.* **278**:27688-27694, 2003.

[23] Russell, D.W. The enzymes, regulation, and genetics of bile acid synthesis. *Annu Rev Biochem.* **72**:137-174, 2003.

[24] Lu, T.T., Repa, J.J., Mangelsdorf, D.J. Orphan nuclear receptors as eLiXiRs and FiXeRs of sterol metabolism. *J Biol Chem*. **276**:37735-37738, 2001.

[25] Francis, G.A., Fayard, E., Picard, F., Auwerx, J. Nuclear receptors and the control of metabolism. *Annu. Rev. Physiol*. **65**: 261-311, 2003.

[26] Edwards, P.A., Kast, H.R., Anisfeld, A.M. BAREing it all: the adoption of LXR and FXR and their roles in lipid homeostasis. *J. Lip Res*. **43**:2-12, 2002.

[27] Murthy, S., Born, E., Mathur, S.N., Field, F.J. LXR/RXR activation enhances basolateral efflux of cholesterol in CaCo-2 cells. *J Lipid Res*. **43**:1054-1564, 2002.

[28] Akiyama, T.E., Sakai, S., Lambert, G., Nicol, C.J., Matsusue, K., Pimprale, S., Lee, Y.H., Ricote, M., Glass, C.K., Brewer, H.B Jr., Gonzalez, F.J. Conditional disruption of the peroxisome proliferator-activated receptor gamma gene in mice results in lowered expression of ABCA1, ABCG1, and apoE in macrophages and reduced cholesterol efflux. *Mol Cell Biol*. **22**:2607-2619, 2002.

[29] Klopotek, A., Hirche, F., Eder, K. PPAR gamma ligand troglitazone lowers cholesterol synthesis in HepG2 and Caco-2 cells via a reduced concentration of nuclear SREBP-2. *Exp Biol Med* (Maywood). **231**:1365-1372, 2006.

[30] McKenna, N.J., Bert, W., O'Malley, B.W. Nuclear receptors and coregulators: an introduction. *NURSA e-Journal*. 2003. Vol. 1, No. 1, ID# 3.06172003.1

[31] Terasaka, N., Hiroshima, A., Koieyama, T., Ubukata, N., Morikawa, Y., Nakai, D., Inaba, T. T-0901317, a synthetic liver X receptor ligand, inhibits development of atherosclerosis in LDL receptor-deficient mice. *FEBS Lett*. **536**:6-11, 2003.

[32] Joseph, S.B., McKilligin, E., Pei, L., Watson, MA., Collins, A.R., Laffitte, B.A., Chen, M., Noh, G., Goodman, J., Hagger, G.N., Tran, J., Tippin, T.K., Wang, X., Lusis, A.J., Hsueh, W.A., Law, R.E., Collins, J.L., Willson, T.M., Tontonoz, P. Synthetic LXR ligand inhibits the development of atherosclerosis in mice. *Proc Natl Acad Sci U S A*. **99**:7604-7609, 2002.

[33] Levin, N., Bischoff, E.D., Daige, C.L., Thomas, D., Vu, C.T., Heyman, R.A., Tangirala, R.K., Schulman, I.G. Macrophage liver X receptor is required for antiatherogenic activity of LXR agonists. *Arterioscler Thromb Vasc Biol*. **25**:135-42, 2005.

[34] Gouni-Berthold, I., Krone, W. Peroxisome proliferator-activated receptor alpha (PPARα) and athero-sclerosis. *Curr Drug Targets Cardiovasc Haematol Disord*. **5**:513-23, 2005.

[35] Lalloyer, F., Fievet, C., Lestavel, S., Torpier, G., van der Veen, J., Touche, V., Bultel, S., Yous, S., Kuipers, F., Paumelle, R., Fruchart, J.C., Staels, B., Tailleux, A. The RXR agonist bexarotene improves cholesterol homeostasis and inhibits atherosclerosis progression in a mouse model of mixed dyslipidemia. *Arterioscler Thromb Vasc Biol*. **26**:2731-7, 2006.

[36] Grover, G.J., Mellstrom, K., Malm, J. Development of the thyroid hormone receptor beta-subtype agonist KB-141: a strategy for body weight reduction and lipid lowering with minimal cardiac side effects. *Cardiovasc Drug Rev*. **23**:133-148, 2005.

[37] Johansson, L., Rudling, M., Scanlan, T.S., Lundasen, T., Webb, P., Baxter, J., Angelin, B., Parini, P. Selective thyroid receptor modulation by GC-1 reduces serum lipids and stimulates steps of reverse cholesterol transport in euthyroid mice. *Proc Natl Acad Sci U S A*. **102**:10297-10302, 2005..

[38] Webb, P. Selective activators of thyroid hormone receptors. *Expert. Opin. Investig. Drugs*. **13**: 489-500, 2004.

[39] Kraiem, Z. Selective agonists and antagonists to thyroid hormone action. *Thyroid*. **15**: 336-339, 2005.

[40] McLaurin, J., Darabi,e A.A., Morrison, M.R. Cholesterol, a Modulator of Membrane-associated Abeta-fibrillogenesis. *Pharmacopsychiatry* **36** (Suppl 2):130-135, 2003.

[41] Yanagisawa, K. Cholesterol and pathological processes in Alzheimer's disease. *J. Neurosci. Res*. **70**: 361-366, 2002.

[42] Simons, M., Keller, P., Dichgans, J., Schulz, J.B. Cholesterol and Alzheimer's disease: is there a link? *Neurology* **57**:1089-1093, 2001.

[43] Tan, Z.S., Seshadri, S., Beiser, A., Wilson, P.W., Kiel, D.P., Tocco, M., D'Agostino, R.B., Wolf, P.A. Plasma total cholesterol level as a risk factor for Alzheimer disease: the Framingham Study. *Arch Intern Med*. **163**:1053-1057, 2003.

[44] Solomon, A., Kareholt, I., Ngandu, T., Winblad, B., Nissinen, A., Tuomilehto, J., Soininen, H., Kivipelto, M. Serum cholesterol changes after midlife and late-life cognition: twenty-one-year follow-up study. *Neurology*. **68**:751-756, 2007.

[45] Kivipelto, M., Helkala, E.L., Laakso, M.P., Hänninen, T., Hallikainen, M., Alhainen, K., Soininen, H., Tuomilehto, J., Nissinen, A. Midlife vascular risk factors and Alzheimer's disease in later life: longitudinal, population based study. *BMJ*. **322**:1447-1451, 2001.

[46] Mielke, M.M., Zandi, P.P., Sjogren, M., Gustafson, D., Ostling, S., Steen, B., Skoog, I. High total cholesterol levels in late life associated with a reduced risk of dementia. *Neurology*. **64**:1689-1695, 2005.

[47] Refolo, L.M., Malester, B., LaFrancois, J., Bryant-Thomas, T., Wang, R., Tint, G.S., Sambamurti, K., Duff, K., Pappolla, M.A. Hypercholesterolemia accelerates the Alzheimer's amyloid pathology in a transgenic mouse model. *Neurobiol Dis*. **7**:321-331, 2000.

[48] Refolo, L.M., Pappolla, M.A., LaFrancois, J., Malester, B., Schmidt, S.D, Thomas-Bryant, T., Tint, G.S., Wang, R., Mercken, M., Petanceska, S.S., Duff, K.E. A cholesterol-lowering drug reduces beta-amyloid pathology in a transgenic mouse model of Alzheimer's disease. *Neurobiol Dis*. **8**:890-899, 2001.

[49] Wolozin, B., Kellman, W., Ruosseau, P., Celesia, G.G., Siegel, G. Decreased prevalence of Alzheimer disease associated with 3-hydroxy-3-methyglutaryl coenzyme A reductase inhibitors. *Arch Neurol*. **57**:1439-1443, 2000.

[50] Canevari, L., Clark, J.B. Alzheimer's disease and cholesterol: the fat connection. *Neurochem Res*. **32**:739-750, 2007.

[51] Zigman, W.B., Schupf, N., Jenkins, E.C., Urv, T.K., Tycko, B., Silverman, W. Cholesterol level, statin use and Alzheimer's disease in adults with Down syndrome. *Neurosci Lett*. **416**:279-284, 2007.

[52] Porcellini, E., Calabrese, E., Guerini, F., Govoni, M., Chiappelli, M., Tumini, E., Morgan, K., Chappell, S., Kalsheker, N., Franceschi, M., Licastro, F. The hydroxy-methyl-glutaryl CoA reductase promoter polymorphism is associated with Alzheimer's risk and cognitive deterioration. *Neurosci Lett*. **416**:66-70, 2007.

[53] Sjogren, M., Gustafsson, K., Syversen, S., Olsson, A., Edman, A., Davidsson, P., Wallin, A., Blennow, K. Treatment with simvastatin in patients with Alzheimer's disease lowers both alpha- and beta-cleaved amyloid precursor protein. *Dement Geriatr Cogn Disord*. **16**:25-30, 2003.

[54] Friedhoff, L.T., Cullen, E.I., Geoghagen, N.S., Buxbaum, J.D. Treatment with controlled-release lovastatin decreases serum concentrations of human beta-amyloid (Abeta) peptide. *Int J Neuropsychopharmacol*. **4**:127-130, 2001.

[55] Kojro, E., Gimpl, G., Lammich, S., Marz, W., Fahrenholz, F. Low cholesterol stimulates the nonamyloidogenic pathway by its effect on the alpha -secretase ADAM 10. *Proc Natl Acad Sci U S A*. **98**:5815-5820, 2001.

[56] Winblad, B., Jelic, V., Kershaw, P., Amatniek, J. Effects of statins on cognitive function in patients with Alzheimer's disease in galantamine clinical trials. *Drugs Aging*. **24**:57-61, 2007.

[57] Won, J.S., Im, Y.B., Khan, M., Contreras, M., Singh, A.K., Singh, I. Lovastatin inhibitis amyloid precursor protein (APP) beta-cleavage through reduction of APP distribution in Lubrol WX extractable low density lipid rafts. *J Neurochem*. 2008 Feb 5; [Epub ahead of print].

[58] Li, L., Cao, D., Kim, H., Lester, R., Fukuchi, K. Simvastatin enhances learning and memory independent of amyloid load in mice. *Ann Neurol*. **60**:729-739, 2006.

[59] Ghribi, O., Larsen, B., Schrag, M., Herman, M.M. High cholesterol content in neurons increases BACE, beta-amyloid, and phosphorylated tau levels in rabbit hippocampus. *Exp Neurol*. **200**:460-467, 2006.

[60] Olesen, O.F., Dagø, L. High density lipoprotein inhibits assembly of amyloid beta-peptides into fibrils. *Biochem Biophys Res Commun*. **270**:62-66, 2000.

[61] Wilson, L.M., Pham, C.L., Jenkins, A.J., Wade, J.D., Hill, A.F., Perugini, M.A., Howlett, G.J. High density lipoproteins bind Abeta and apolipoprotein C-II amyloid fibrils. *J Lipid Res*. **47**:755-760, 2006.

[62] Simons, M., Keller, P., De Strooper, B., Beyreuther, K., Dotti, C.G., Simons, K. Cholesterol depletion inhibits the generation of beta-amyloid in hippocampal neurons. *Proc Natl Acad Sci U S A*. **95**:6460-6464, 1998.

[63] Koudinov, A.R., Koudinova, N.V. Essential role for cholesterol in synaptic plasticity and neural degeneration. *FASEB J*. **15**: 1858-1860, 2001.

[64] Koudinov, A.R., Koudinova, N.V. Cholesterol, synaptic function and Alzheimer's disease. *Pharmacopsychiatry*. **36** Suppl 2:S107-112, 2003.

[65] Ji, S-R., Wu, Y., Sui, S-F. Cholesterol is an important factor affecting the membrane insertion of β-amyloid peptide (Aβ1-40), which may potentially inhibit the fibril formation. *J. Biol. Chem*. **277**: 6273-6279, 2002.

[66] Wahrle, S.E., Jiang, H., Parsadanian, M., Kim, J., Li, A., Knoten, A., Jain, S., Hirsch-Reinshagen, V., Wellington, C.L, Bales, K.R., Paul, S.M., Holtzman, D.M. Overexpression of ABCA1 reduces amyloid deposition in the PDAPP mouse model of Alzheimer disease. *J Clin Invest*. **118**:671-682, 2008.

[67] Wahrle, S.E., Jiang, H., Parsadanian, M., Hartman, R.E., Bales, K.R., Paul, S.M., Holtzman, D.M. Deletion of Abca1 increases Abeta deposition in the PDAPP transgenic mouse model of Alzheimer disease. *J Biol Chem*. **280**:43236-43242, 2005.

[68] Wollmer, M.A., Streffer, J.R., Lütjohann, D., Tsolaki, M., Iakovidou, V., Hegi, T., Pasch, T., Jung, H.H., Bergmann, K., Nitsch, R.M., Hock, C., Papassotiropoulos, A. ABCA1 modulates CSF cholesterol levels and influences the age at onset of Alzheimer's disease. *Neurobiol Aging*. **24**:421-426, 2003.

[69] Katzov, H., Chalmers, K., Palmgren, J., Andreasen, N., Johansson, B., Cairns, N.J., Gatz, M., Wilcock, G.K., Love, S., Pedersen, N.L., Brookes, A.J., Blennow, K., Kehoe,

P.G., Prince, J.A. Genetic variants of ABCA1 modify Alzheimer disease risk and quantitative traits related to beta-amyloid metabolism. *Hum Mutat.* **23**:358-367, 2004.

[70] Wavrant-De Vrièze, F., Compton, D., Womick, M., Arepalli, S., Adighib,e O., Li, L., Pérez-Tur, J., Hardy, J. ABCA1 polymorphisms and Alzheimer's disease. *Neurosci Lett.* **416**:180-183, 2007.

[71] Tansley, G.H., Burgess, B.L., Bryan, M.T., Su, Y., Hirsch-Reinshagen, V., Pearce, J., Chan, J.Y., Wilkinson, A., Evans, J., Naus, K.E., McIsaac, S., Bromley, K., Song, W., Yang, H.C., Wang, N., DeMattos, R.B., Wellington, C.L. The cholesterol transporter ABCG1 modulates the subcellular distribution and proteolytic processing of beta-amyloid precursor protein. *J Lipid Res.* **48**:1022-1034, 2007.

[72] Wollmer, M.A., Sleegers, K., Ingelsson, M., Zekanowski, C., Brouwers, N., Maruszak, A., Brunner, F., Huynh, K.D., Kilander, L., Brundin, R.M., Hedlund, M., Giedraitis, V., Glaser, A., Engelborghs, S., De Deyn, P.P., Kapaki, E., Tsolaki. M., Daniilidou, M., Molyva, D., Paraskevas, G.P., Thal, D.R., Barcikowska, M., Kuznicki, J., Lannfelt, L., Van Broeckhoven, C., Nitsch, R.M., Hock, C., Papassotiropoulos, A. Association study of cholesterol-related genes in Alzheimer's disease. *Neurogenetics.* **8**:179-188, 2007.

[73] Hur, J.Y., Welander, H., Behbahani, H., Aoki, M., Frånberg, J., Winblad, B., Frykman, S., Tjernberg, L.O. Active gamma-secretase is localized to detergent-resistant membranes in human brain. *FEBS J.* 2008 Feb 6; [Epub ahead of print].

[74] Bodovitz, S., Klein, W.L. Cholesterol modulates alpha-secretase cleavage of amyloid precursor protein. *J Biol Chem.* **271**:4436-4440, 1996.

[75] Kojro, E., Gimpl, G., Lammich, S., Marz, W., Fahrenholz, F. Low cholesterol stimulates the nonamyloidogenic pathway by its effect on the alpha -secretase ADAM 10. *Proc Natl Acad Sci U S A.* **98**:5815-5820, 2001.

[76] Parvathy, S., Ehrlich, M., Pedrini, S., Diaz, N., Refolo, L., Buxbaum, J.D,, Bogush, A., Petanceska, S., Gandy, S. Atorvastatin-induced activation of Alzheimer's alpha secretase is resistant to standard inhibitors of protein phosphorylation-regulated ectodomain shedding. *J Neurochem.* **90**:1005-10, 2004.

[77] Famer, D., Meaney, S., Mousavi, M., Nordberg, A., Björkhem, I., Crisby, M.. Regulation of alpha- and beta-secretase activity by oxysterols: cerebrosterol stimulates processing of APP via the alpha-secretase pathway. *Biochem Biophys Res Commun.* **359**:46-50, 2007.

[78] Cordy, J.M., Hooper, N.M., Turner, A.J. Exclusively targeting beta-secretase to lipid rafts by GPI-anchor addition up-regulates beta-site processing of the amyloid precursor protein. *Proc Natl Acad Sci U S A.* **100**:11735-11740, 2003.

[79] Kalvodova, L., Kahya, N., Schwille, P., Ehehalt, R., Verkade, P., Drechsel, D., Simons, K. Lipids as modulators of proteolytic activity of BACE: involvement of cholesterol, glycosphingolipids, and anionic phospholipids in vitro. *J Biol Chem.* **280**:36815-36823, 2005.

[80] Wahrle, S., Das, P., Nyborg, A.C., McLendon, C., Shoji, M., Kawarabayashi, T., Younkin, L.H., Younkin, S.G., Golde, T.E. Cholesterol-dependent gamma-secretase activity in buoyant cholesterol-rich membrane microdomains. *Neurobiol Dis.* **9**:11-23, 2002.

[81] Koldamova, R.P., Lefterov, I.M., Staufenbiel, M., Wolfe, D., Huang, S., Glorioso. J.C., Walter, M., Roth, M.G., Lazo, J.S. The liver X receptor ligand T0901317 decreases

amyloid beta production in vitro and in a mouse model of Alzheimer's disease. *J Biol Chem.* **280**:4079-4088, 2005.

[82] Xiong, H., Callaghan, D., Jones, A., Walker, D.G., Lue, L-H., Beach, T.G., Sue, L.I., Xu, H., Stanimirovic, D., Zhang, W. Cholesterol retention in Alzheimer's brain is responsible for high β-and γ-secretase activities and Aβ production. *Neurobiol. Dis.* **29**: 422-437, 2008.

[83] Huang, Y. Molecular and cellular mechanisms of apolipoprotein E4 neurotoxicity and potential therapeutic strategies. *Curr. Opin. Drug Discov. Devel.* **9**: 627-641, 2006.

[84] Packard, C.J., Westendorp, R.G., Stott, D.J., Caslake, M.J., Murray, H.M., Shepherd, J., Blauw, G.J., Murphy, M.B., Bollen, E.L., Buckley, B.M., Cobbe, S.M., Ford, I., Gaw, A., Hyland, M., Jukema, J.W., Kamper, A.M., Macfarlane, P.W., Jolles, J., Perry, I.J., Sweeney, B.J., Twomey, C., Prospective Study of Pravastatin in the Elderly at Risk Group. Association between apolipoprotein E4 and cognitive decline in elderly adults. *J Am Geriatr Soc.* **55**:1777-1785, 2007.

[85] Bandaru, V.V., Troncoso, J., Wheeler, D., Pletnikova, O., Wang, J., Conant, K., Haughey, N.J. ApoE4 disrupts sterol and sphingolipid metabolism in Alzheimer's but not normal brain. *Neurobiol Aging.* 2007 Sep 19; [Epub ahead of print].

[86] Ye, S., Huang, Y., Müllendorff, K., Dong, L., Giedt, G., Meng, E.C., Cohen, F.E., Kuntz, I.D., Weisgraber, K.H., Mahley, R.W. Apolipoprotein (apo) E4 enhances amyloid beta peptide production in cultured neuronal cells: apoE structure as a potential therapeutic target. *Proc Natl Acad Sci U S A.* **102**: 18700-18705, 2005.

INDEX

A

B

C

D

E

F

G

H

I

M

N

O

P

Q

R

S

T

U

V

W

X

Y

Z